Assessing Working Conditions
– The European practice –

EF/96/06/EN

Assessing Working Conditions
– The European practice –

by

J.C.M. Mossink
H.G. De Gier

TNO Prevention and Health
Leiden
The Netherlands

 European Foundation for the
Improvement of Living and Working Conditions
Loughlinstown, Dublin 18, Ireland
Tel: (+353) 1 282 6888 Fax: (+353) 1 282 6456

Cataloguing data can be found at the end of this publication

Luxembourg: Office for Official Publications of the European Communities, 1996

ISBN 92-827-6098-7

© European Foundation for the Improvement of Living and Working Conditions, 1996

For rights of translation or reproduction, applications should be made to the Director, European Foundation for the Improvement of Living and Working Conditions, Loughlinstown, Dublin 18, Ireland.

Printed in Ireland

Foreword

In 1987 the European Foundation started a programme of work aimed at improving the monitoring of working conditions in the European Union, in particular with regard to occupational health.

Among other projects a data bank (HASTE) describing the various systems set up in all member states to monitor working conditions was created and the First European questionnaire-based survey was carried out in 1991 among 13000 workers.

While these activities were welcomed, suggestions were made to the Foundation that working conditions should not only be monitored at the macro-economic level but also at the company level. It is in fact essential that companies are in a position to make a precise analysis of the working conditions of their staff, whether the reasons be social (improvement of the existing environment), technical (design of new production systems and workplaces), economical (quality), legal, etc.

This led the Foundation to set up a working group to collect and consolidate examples of best practice throughout Europe in order to help companies to set up assessment policies, carry out the assessments and finally to make best use of the assessment findings.

In the meantime the 1989 European Framework Directive on safety and health at work was issued, requiring companies to carry out systematic assessment of the risks to the health and safety of their workers.

The Foundation wishes to stress that the recommendations set out in the present document are in no way legally related to the above mentioned Directive. The present report does not prescribe but indicates what can be done. Its scope is also broader than the specific issue of 'risk assessment'. This being said, it could contribute to helping companies to meet their new legal obligations and thus complement the Framework Directive.

The present document, which was drafted by Jos Mossink and Erik de Gier from TNO, is the result of the collective work carried out by a group which included:
- Emilio Castejón, INSHT, Barcelona;
- Dagmar Diergarten, BDA, Köln;
- Erik de Gier, TNO-PG, Leiden;
- Burkhard Hoffmann, BGZ, Sankt Augustin;
- Karl Kuhn, Bundesanstalt für Arbeitsschutz, Dortmund;
- Jos Mossink, TNO-PG, Leiden;
- Jens Øland, BST Storstrøm Nord, Næstved;
- Evelyne Polzhuber, ANACT, Montrouge;
- Francesco Violante, Occupational Diseases Diagnostic Centre, Bologna;
- Laurent Vogel, TUTB, Brussels;
- Françoise Doppler, Aerospatiale, Paris;
- Jaume Costa, European Foundation, Dublin;
- Claudio Stanzani, Sindnova-CISL, Rome;
- Horst Kloppenburg, Commis. of the European Communities, DG 5. Luxembourg;
- Pascal Paoli, European Foundation, Dublin.

The work was coordinated by Pascal Paoli.

The Foundation wishes to express its thanks to all those who contributed to the present project.

Table of Contents

1.	INTRODUCTION	13
	1.1 Background	13
	1.2 Workplace assessment	14
	1.3 Purpose of this book	14
	1.4 Structure of the book	16
2.	BENEFITS OF WORKPLACE ASSESSMENT	19
	2.1 Introduction	19
	2.2 Legal context	21
	2.3 Costs of workplace assessment	22
	2.4 Benefits for companies	23
	2.5 Additional benefits for workers	28
3.	BASICS AND FUNDAMENTALS	31
	3.1 Introduction	31
	3.2 What is workplace assessment?	32
	3.3 National context	34
	3.4 Goals	35
	3.5 Types of assessments	36
4.	GETTING STARTED WITH WORKPLACE ASSESSMENT	49
	4.1 Introduction	49
	4.2 Selecting goals and specifying outcomes	50
	4.3 Persons involved	52
	4.4 Workers' participation	55
	4.5 Selecting scope and subjects	62
	4.6 Setting up criteria	63
	4.7 Planning	66
	4.8 Follow-up	66

5.		**CARRYING OUT WORKPLACE ASSESSMENTS**	67
	5.1	Introduction	67
	5.2	Collecting information	67
	5.3	Evaluating results	78
	5.4	Selection of priorities	83
	5.5	Action planning	90
6.		**IMPROVING PERFORMANCE**	93
	6.1	Introduction	93
	6.2	Measurement and review	94
	6.3	The management cycle	98
	6.4	Introducing safety and health management	102
	6.5	Investigating performance: the audit	106
7.		**SUMMARY AND GUIDELINES**	109
8.		**GLOSSARY**	113
9.		**REFERENCES AND FURTHER READING**	117

LIST OF FIGURES

Figure 1.1	Structure of the book.	17
Figure 2.1	Factors which prompted health actions. Results from a survey in seven EU countries.	22
Figure 2.2	The benefits of health actions. Results from a survey in seven EU countries.	24
Figure 2.3	Spectrum of occupational safety and health.	29
Figure 3.1	Inspection of working situations according to the cascade approach: Grundmethodik Gefährdungsanalyse, general overview of the instrument and visualisation of the structure.	37
Figure 3.2	Inspection of working situations according to the cascade approach: Grundmethodik Gefährdungsanalyse, global checklist.	38
Figure 3.3	Inspection of working situations according to the cascade approach: Grundmethodik Gefährdungsanalyse, detailed structure.	39
Figure 3.4	Importance of knowledge about real work in reference situations.	41
Figure 3.5	Knowledge of reference situations in the context of a design project.	42
Figure 3.6	Model for noise assessment, starting from three different situations.	43
Figure 3.7	Assessment of stress problems, part of a checklist.	44
Figure 3.8	Strategic assessments. Part of the method for the evaluation of working conditions.	46
Figure 4.1	Schematic outline of a workplace assessment process.	51
Figure 4.2	Organization of activities for workplace assessment according to the 'internal control' principle.	53
Figure 4.3	Workgroups in redesign project at 'Le Monde'.	58
Figure 4.4	Part of the "do it yourself" checklist for the evaluation of administrative work at computer screens.	59
Figure 4.5	Possible structure of the workplace assessment process.	64
Figure 5.1	Part of an ergonomic observation sheet suited for tasks with a cycle time of less than 6 minutes.	70

Figure 5.2 Closed checklist in order to find lighting problems. 72

Figure 5.3 Safety and health circles at Asea Brown Boveri. 73

Figure 5.4 Part of the Volkswagen questionnaire on health and wellbeing. 75

Figure 5.5 The projects in the programme of the European Foundation for the Improvement of Living and Working Conditions relating to health and safety at work. 76

Figure 5.6 Classification of risks and selection of actions according to the German DIN 19250 standard. 82

Figure 5.7 Relative ranking techniques for prioritizing risks. Sequence for determining whether action is required. 86

Figure 5.8 Costs - effects analysis of possible measures as a tool in action planning. 90

Figure 5.9 Sample action plan. 90

Figure 6.1 Overview of a risk management system. 94

Figure 6.2 Four step safety management system. 96

Figure 6.3 Three level safety and health management system. 97

Figure 6.4 Possible options for organizing integrated quality, environmental and safety and health control. 102

Figure 6.5 Audit as a part of safety and health management systems. 104

LIST OF TABLES

Table 3.1 Differences between workplace assessment and risk assessment. 33

Table 3.2 Overview of some workplace assessment methods. 40

Table 3.3 Some instruments for thematic assessments. 42

Table 3.4 Selection aid for assessment types. 47

Table 4.1 Overview of possible persons or groups involved, tasks and responsibilities. 54

Table 4.2 Overview of practices and comments on worker participation. 60

Table 4.3 Overview of models for the planning of workplace assessment. 64

Table 5.1 Examples of databases and catalogues. 77

Table 5.2 Possible sources of criteria. 79

Table 5.3	Overview of risk estimation techniques. Both quantitative and semi-quantitative techniques are summarized.	81
Table 5.4	Injury/illness-cause model.	87
Table 6.1	Overview of audit instruments for safety.	104

LIST OF EXAMPLES

Example 2.1	Motives for management support of occupational safety and health.	20
Example 2.2	Obligations according to the Council Directive.	21
Example 2.3	Working conditions in the international standard for quality management and quality system elements (ISO 9004).	25
Example 2.4	Safety pays.	26
Example 2.5	Beneficial effects of lower absenteeism on company performance.	27
Example 3.1	Occasions on which workplace assessments may be started.	35
Example 4.1	Overview of activities to perform at the start of a workplace assessment process.	51
Example 4.2	Workgroups as a means of participation.	58
Example 4.3	Eleven-step approach to planning workplace assessments.	65
Example 5.1	Workload assessment in the chemical industry; workplace assessment at DAF.	69
Example 5.2	Health circles at a German steel mill.	74
Example 5.3	Company safety and health reports (Germany).	84
Example 6.1	Objectives of performance standards.	95
Example 6.2	Tasks and responsibilities with respect to internal control of safety and health in Norway.	99
Example 6.3	Total Safety Management at ABB.	100
Example 6.4	Organization and structure of combined quality, occupational safety and health, environment and organization management system.	103
Example 6.5	Overview of the contents of the OSART questionnaire.	105

Introduction 1

1.1 Background

Interest in working conditions and occupational safety and health is growing. More and more it is being recognized that good working conditions are an essential part of management. Modern practices of human resources management pay particular attention to working conditions, of which job satisfaction is one aspect.

Good working conditions are important in the prevention of accidents and occupational diseases. In this way a contribution is made to general health improvement and also to the reduction of costs on a national, branch and company level. In general, workplace assessment can contribute to the quality of company management, for instance by enhancing communication. There may be positive effects on many aspects of organizational performance.

Individual companies benefit from better working conditions in a number of ways, such as: less absenteeism, higher productivity and better quality of products and services. For the improvement of working conditions, assessment of the actual situation is the starting point.

In this book the concept of workplace assessment (WPA) will be worked out in the European context. Best practices and methods from countries of the European Union have been collected.

1.2 Workplace assessment

Workplace assessment is a systematic investigation of work, in all its aspects, in order to find situations or activities that may cause undesirable effects such as accidents, diseases or discomfort. The evaluation of adverse situations is also part of the assessment.

Workplace assessment is a broad concept, in which all aspects of work are incorporated. Attention is paid to, for instance, physical and chemical work environment, ergonomics, mental and physical workload. Organizational aspects and wellbeing at work are also considered.

Workplace assessment can have many functions within the company. It can investigate working conditions, fulfil legal obligations or start an improvement process in workplaces. Workplace assessments are the starting point of a process consisting of the design and implementation of improvements for working conditions. Positive results such as improved productivity can also be obtained.

1.3 Purpose of this book

This book is an introduction to the subject of "workplace assessment". It is a concise overview of methods, techniques and instruments for making assessments of working conditions in firms. As such, it provides ideas as to how workplace assessment can be incorporated in company policies.
The book is especially meant for companies or organizations that want to begin workplace assessment. Companies who already have experience in this field can also benefit from this publication, as the topic is illustrated by the best methods and practices used in Europe. Some different views on investigating working conditions are also discussed.

Good working conditions and optimal safety are primarily the concern of company management. Occupational safety and health is one aspect of management responsibilities, just as much as profitability, quality or the environment. Therefore this book is written primarily for those people who are in some way responsible for working conditions. These are:

Top management
Top management must be the driving force in innovation stimulating the health of a company. In this book the profitability of workplace assessment is demonstrated by illustrations from best practice in Europe.

Middle and lower management
The daily routine, the operational aspects of production, are the responsibility of middle management. As a result, the practice of workplace assessment has to be stimulated and controlled by line management. Working conditions deserve attention just as much as organizational matters or production management. Carrying out workplace assessments, in cooperation with workers and possibly with the support of experts, is one of their tasks. For middle and lower management this book gives a wealth of information on organizing and performing workplace assessments.

Workers and their representatives
In some situations, managers will not carry out complete workplace assessments themselves, but leave part of it to workers or experts. For instance the input of information and the generation of ideas for improvements are activities that workers should participate in.
This book will provide information to them on the way participative workplace assessments are structured, and can help them to create participative structures in workplace assessment in their own company. The role of more formal participative structures (such as works councils, trade unions and appointed safety and health delegates) is also dealt with.

Experts, working conditions consultants
Until now, workplace assessment and occupational safety and health have been very much the domain of experts. For effective safety and health policies, however, it is necessary to shift responsibility to line management

and give experts a supportive role. As a consequence this book is not primarily meant for experts. Nevertheless the contents may add something to their own knowledge and experience.

1.4 Structure of the book

There is no single best way to perform workplace assessments. In general, ready-made methods which are imposed on companies do not produce good results. It is preferable to have people within the company set up their own workplace assessments and work according to their own tradition. Therefore, this book is an open document. The aim is to give an overview and explanation of several ways to perform a workplace assessment. It is a toolkit, a collection of methods and practices. The common view on workplace assessment is consistent, but there is no one single method promoted. This is consistent with our strong belief that every company should adopt or create a workplace assessment practice that fits the company culture and its typical working condition problems.

Guidance for reading

If some members of the organization still have to be convinced, start reading at **Chapter 2**, where a number of benefits from workplace assessments are described. It is shown that there is a number of good reasons to start with workplace assessment.

If the concept of workplace assessment is new to the company, but there is a positive attitude, **Chapter 3** is a good start. Some basic concepts of workplace assessment are clarified, and an overview of possible strategies and general methods is given.

Once familiar with the concept of workplace assessment, a start can be made. For those with little experience in setting up a workplace assessment, **Chapter 4**, 'Getting Started' will offer some practical guidelines and possibilities.

Chapter 5 gives more details on the technical side of workplace assessment. It describes how one can proceed from a first inventory of working condition problems to an action plan with solutions that will work. In this chapter techniques and methods, as well as company practices, are described.

When a company has developed a practice of workplace assessment, a need to obtain maximum efficiency may emerge. **Chapter 6** will go into the ways in which the efficiency of workplace assessment can be improved. Attention is given to the management of occupational safety and health.
Chapter 7 describes an overview of workplace assessment. In this summarizing chapter, some guidelines can be found as well.

The guidance for reading is also illustrated in figure 1.1.

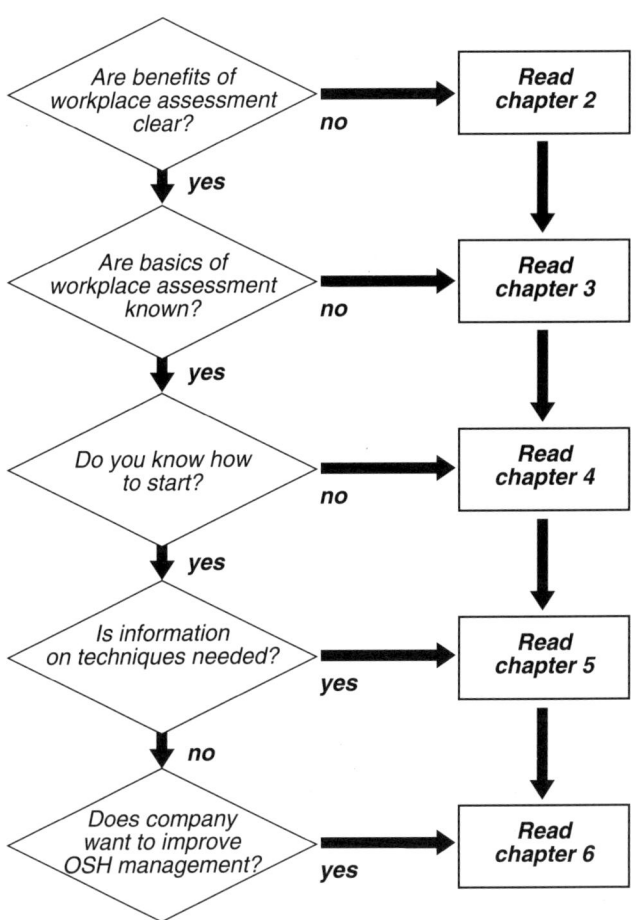

Figure 1.1   Structure of the book.

Benefits of Workplace Assessment

2.1 Introduction

There are many motives for paying attention to working conditions within companies. The economic context of actions is almost always considered. Legal obligations also play a role. In practice more ethical motives, such as having a healthy company, will become more important (also see example 2.1).

Workers and the company as a whole have the same interests with respect to working conditions. For instance, high motivation and job satisfaction benefit both workers and the company as a whole. Good economic performance gives more job security and better opportunities for improvement of working conditions.

Nevertheless some benefits are especially of interest to workers (e.g. fewer health risks), whilst some others are primarily applicable to the company (e.g. higher production, improved flexibility).

Example 2.1 *Motives for management support of occupational safety and health.*

In Denmark a study was performed in order to find out why legal obligations did not lead to widespread occupational safety and health programmes within companies. One of the issues was what motivated top management. Mostly a mixture of motives was behind management support.

Motives mentioned most were:
1. general attitude;
2. ambition to master this part of management practice;
3. ambition of the firm to have a good image, also from a working conditions perspective;
4. endeavouring to reduce risk (in high-risk companies);
5. reduction of costs due to sick leave;
6. inspiration of external consultants;
7. pressure from the Labour Inspectorate.

(Source: Jensen et al., 1993)

From the motives mentioned in example 2.1 it can be concluded that occupational safety and health programmes are most strongly initiated when the company has a positive attitude. This attitude arises when managers think occupational safety and health is good for the company. That means economic and organizational benefits must be perceived.

Workplace assessment contributes to a number of benefits which may be of an economic, social or human nature. Examples at a company level are:
- prevention of diseases and accidents resulting in reduced costs;
- reduction of sick leave and lower personnel turnover;
- improvement of communication, consent on the topic of working conditions;
- commitment of workers and improvement of industrial relations;
- enhancement of quality;
- improvement of productivity and efficiency;
- better position on the labour market, more attractive jobs.

Over time, personnel tend to make higher demands on their workplaces, and working conditions become more important to them. In some specialized professions in particular, working conditions may be decisive in recruiting good personnel.

Some examples of additional benefits for workers are:
- better working conditions;
- participation;
- safer and healthier work.

2.2 Legal context

According to Council Directive 89/391/EC, risk assessment has become a legal obligation in several countries of the European Union. The Directive indicates the aims for legislation. Each Member State then chooses the most appropriate way with implement it by adaptation to its own legislation. Compliance with legislation may be a goal in itself, but in this way working conditions assessment raises costs. On the other hand if the assessment is part of a preventive policy (as the Council Directive requires) more benefits can be gained from assessments.

Example 2.2   Obligations according to the Council Directive.

The council directive of 12 June 1989 on the introduction of measures to encourage improvements in the safety and health of workers at work (89/391/EC) states:

Article 6   General obligations on employers

1   *"[...] The employer shall take the measures necessary for the safety and health of workers [...]"*

2   *"[..] on the basis of the following general principles of prevention:*

   2(g) developing a coherent overall prevention policy which covers technology, organization of work, working conditions, social relationships and the influence of the factors related to the working environment"

3   *(a)"[...] the employer shall [...] evaluate the risks to the safety of workers [...]"*

Article 9   various obligations on employers

1(a)   *"The employer shall [...] be in possession of an assessment of the risks to safety and health at work [...]"*

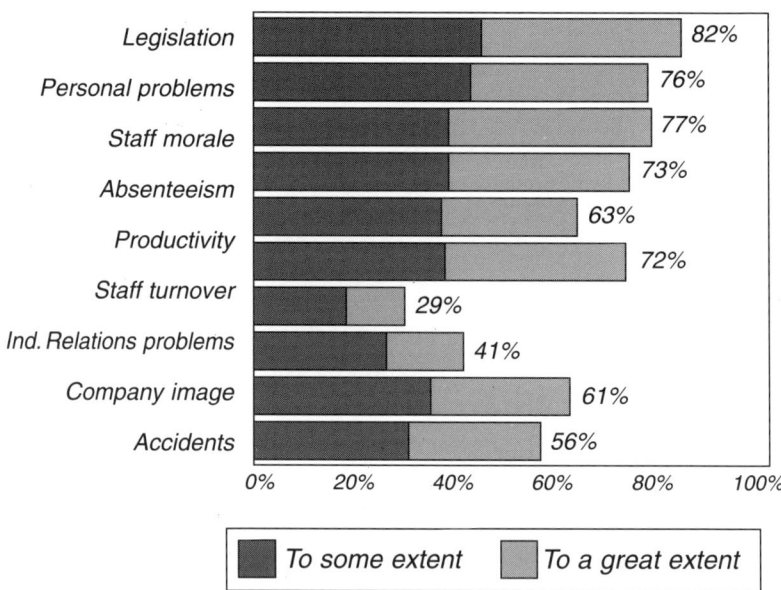

Figure 2.1 Factors which prompted health actions. Results from a survey in seven EU countries. The bar chart shows the percentage of companies that indicated the topics which led to health improvement actions. (Source: Wynne & Clarkin, 1992).

However, to make optimum use of workplace assessments requires additional efforts in properly organizing and conducting assessments. Generally speaking, legal obligations are a poor motive for paying attention to occupational safety and health. Nevertheless, compliance with regulations adds to the total number of benefits which can be obtained from good working conditions. New regulations may be an immediate cause to perform workplace assessments, especially for those companies that have no negative attitude, but never had a good reason to start. In figure 2.1 this is illustrated.

2.3 Costs of workplace assessment

There are virtually no benefits without a prior investment. Although the costs of a good functioning safety and health practice need not be very high, there are some expenses:
- some lost production time due to cooperation of production personnel in assessments, reporting and discussions;

- cost of training and information;
- costs of advising and consulting experts: in some countries companies are obliged to hire experts from health and safety services;
- project costs of design and implementation improvements;
- small administrative overheads (preparing reports and coordination of activities).

It is worthwhile to keep in mind that performing assessments can give rise to expectations among workers and lower management. A budget reservation for improvements is essential in order to maintain the confidence and the cooperation of employees.
Furthermore, workplace assessment is sometimes seen as an additional burden, which must compete with many other calls on the attention of managers.

2.4 Benefits for companies

In general many benefits for companies can be demonstrated. Most of those may not be valid in all situations. A company with low sick leave rates does not profit as much from additional prevention as a company with high absenteeism. Increase in flexibility is an incentive for those companies that operate in a quickly changing market in which fast response to change is required. However, in general a number of positive effects can be obtained, as is illustrated in figure 2.2.
For small and medium-sized companies, benefits are sometimes hard to indicate. With few people employed, sick leave and accident rates give no reliable information.

Quality management
Total Quality Management applies to all aspects of management, including occupational safety and health. Constant high quality can only be achieved with high standards of working conditions. Therefore working conditions are part of the quality standards of the ISO 9000 series.
Elements of the ISO 9004 standard for quality management that are related to working conditions and human factors are listed in example 2.3.

New production concepts
The economic situation favours new concepts of production, lean production being one of them. Companies tend to concentrate on their core

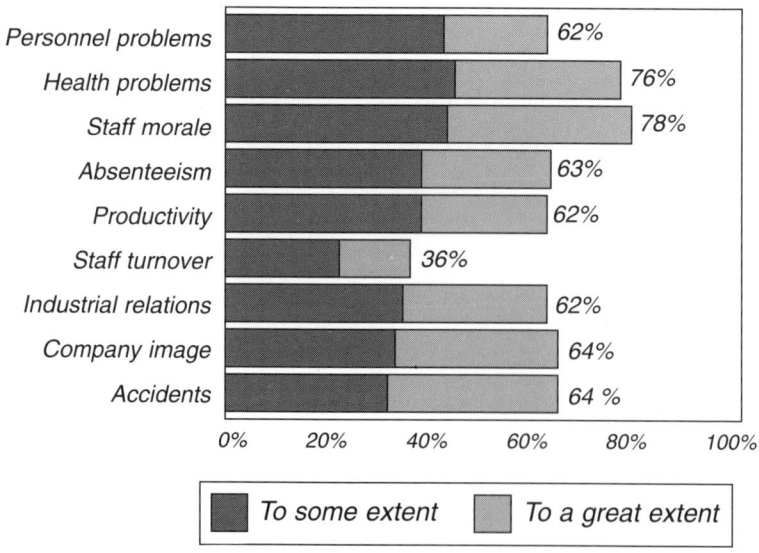

Figure 2.2. *The benefits of health actions. Results from a survey in seven EU countries. The bar chart shows the percentage of companies that indicated a positive effect. (Source: Wynne & Clarkin, 1992).*

activities in order to respond quickly to ever-changing market demands and to reduce costs. Supporting activities are carried out by contractors. Stock is kept to a minimum. The production process is subject to a continual optimization.

Company standards of productivity, efficiency, flexibility and quality keep improving. Workers have to increase their output without concessions to quality. This cannot be achieved adequately without the contribution and cooperation of workers. Good working conditions are also of great importance. So, workplace assessment may have an important role in the improvement process, and some companies may even be forced to pay attention to workplace assessment in order to control occupational safety and health matters that directly relate to production. Some issues of workplace assessment, such as stress risks, participation and working practices, must be addressed. Furthermore, workplace assessment can improve the coordination and transfer of know-how within the company.

> **Example 2.3** **Working conditions in the international standard for quality management and quality system elements (ISO 9004).**
>
> In the International Standard for Quality management and quality system elements (ISO 9004), a number of quidelines for quality control are explained. Some of them directly or indirectly refer to working conditions or human factors.
>
> **0.2 Organizational goals**
> In order to meet its objectives, the company should organize itself in such a way that technical, administrative and human factors affecting the quality of its products and services will be under control [....].
>
> **0.3 Meeting company/customer needs**
> [....] For the company, there is a need to attain and to maintain the desired quality at an optimum cost; the fulfilment of this quality is related to the planned and efficient utilization of the technical, human and material resources available to the company.
>
> **0.4.2.1 Risk considerations for the company**
> Consideration has to be given to risks related to deficient products or services which lead to loss of image or reputation, loss of market, complaints, claims, liability, waste of human and financial resources.
>
> **5.2.4 Resources and personnel**
> Management should provide sufficient and appropriate resources essential to the implementation of quality policies and the achievement of quality objectives. These resources may include:
> a) human resources and specialized skills
> b) design and development equipment
> c) manufacturing equipment
> [....].
>
> **5.4.3 Carrying out the audit**
> Objective evaluations of quality systems by competent personnel may include the following activities or areas:
> a) organizational structures
> b) administrative and operational procedures
> c) personnel, equipment and material resources
> d) work areas, operations and processes
> [....].
>
> **18.3 Motivation (of personnel)**
> Motivation of personnel begins with their understanding of the tasks [....].
>
> (Source: ISO, 1987)

Preservation of resources
Safety and health management contributes to the preservation of resources. It may also result in higher productivity, better quality and improved flexibility. For some companies there is an intangible spin-off as well. In the business-to-business market, companies must perform well on all aspects.

> **Example 2.4   Safety pays.**
>
> **Construction industry**
> In the European construction industry in 1988 750 000 accidents and 1 413 deaths were registered. The total cost of accidents was about 15 billion ecu. It was estimated that the cost of accidents was about 3% of the total volume of business. The net profits are about 1.3% of the total business volume. So the cost of accidents is more than double the average profits.
> The costs of preventive policies and accident prevention in construction sites are 1.5% of the total business volume. This is about half the cost of accidents.
>
> (Source: Lorent, 1993)
>
> **Paper industry**
> Paper is produced by shiftwork, using machines of over 100 m long, 12 m high and 10 m wide. The work is very demanding on the staff, especially during scheduled and unforeseen downtimes. Staff are exposed to above average hazards. In 50 companies a management strategy has been tested to prevent accidents.
> In five years a reduction of 2027 accidents was realized (1987: 8560 accidents; 1992: 6533; about 25% reduction). The average absence is 15 days. Every lost working day costs the company about DM 1000. In five years 30.4 million DM was saved.
>
> (Source Meyer et al. 1993)

Accidents and sick leave
Good working conditions and low accident and sick leave rates may give a (small but decisive) competitive advantage. First, the cost of accidents and sick leave may be considerable. Second, low accident rates and low absenteeism are indicators of the overall quality of company management.

Accidents and ill health may result in several costs:
- loss of working days, costs of additional staffing;
- liability insurance costs (several systems are based on accident or illness rates);
- lost production;
- damage to products, machinery and premises;
- non-tangible losses (for instance company image);

In example 2.4 two cases of cost of accidents in relation to prevention costs are described. In these cases it turns out that improving safety would be very cost effective.

Reduction of sick leave may have a number of advantages besides lowering the costs of insurance and replacements. Often, sick leave of workers causes

> **Example 2.5 Beneficial effects of lower absenteeism on company performance.**
>
> **Company profile**
> A medium-sized company produces packaging materials in a highly competitive situation. Strong points of the company are its quality and its just-in-time deliveries. The company was confronted with high sick leave rates.
>
> **Problems due to absenteeism**
> The main problem with sick leave was not so much the direct cost, but the trouble it caused in the planning. The absence of specialized machine operators forced the company to deploy less experienced operators. As these could not produce optimal quality at maximum speed, production rates dropped, there was more reworking and deliveries were late.
>
> **Reduction of sick leave**
> The company had success with a programme to lower absenteeism. There is a wish to stabilize sick leave rates at a low level. The company is now convinced that better working conditions are a prerequisite.
>
> A working conditions committee was founded. Much attention was paid to musculo-skeletal problems and some investments were made to improve some workplaces. However the economic feasibility of further improvements is doubtful.
>
> **Workplace assessment**
> A workplace assessment revealed other possibilities for improvements:
> -some safety hazards were identified;
> -some problems with respect to climate and noise were clearly identified;
> -a number of stress risks and organizational problems were encountered.
> The latter turned out to be important. It was felt that part of the sick leave was caused by organizational factors and stress. Resolving stress problems would also have positive effects on the quality of the planning, and thus contribute to the company's competitive position.
>
> *(Source: TNO-PG practice)*

problems in day-to-day production. Usually, replacements are not able to perform at the required standard at once. An illustration is given in example 2.5.

In some situations, attempts to reduce sick leave rates do not pay off. For instance, when absenteeism is already low, much effort is expended for little gain. Also, in small and medium-sized enterprises, costs may be relatively high.

Design projects
Workplace assessment is especially useful in design or investment projects. The knowledge of work and the work process which is established by a

workplace assessment can lead to a better design. It has been shown that paying attention to ergonomics and working conditions in the early stages of design processes leads to better working conditions, fewer accidents and sick leave and higher productivity and quality at virtually no cost. Companies that frequently change their products or production lines may benefit particularly from design practices in which workplace assessment has a prominent place.

Non-economic benefits

Carrying out workplace assessments can have some benefits that do not have an immediate economic benefit, but do contribute to the overall company performance.
- Better insight is gained into the work process and the problems workers experience in carrying out their everyday work.
- Improvement of decision-making processes concerning changes in production equipment, premises or organizational structure can be made. Some aspects are relevant:
  = more knowledge of how work is actually done;
  = better communication;
  = improved relations;
  = consent about policies and actions.
- Higher job satisfaction and motivation of workers is often a result of paying attention to working conditions.

The concept of ethical management is developing. For some companies ethical management, and the prevention of occupational diseases in particular, has proved to be a powerful marketing tool.

2.5 Additional benefits for workers

Workplace assessment is a topic that directly deals with the everyday routine of workers. For workers the direct benefits are:
- Workplace assessment is a way in which to investigate complaints, because it looks for the causes of problems in working conditions;
- One of the main objectives is to look for solutions for problems and improvements, in order to be able to do the work better and in an easier way.

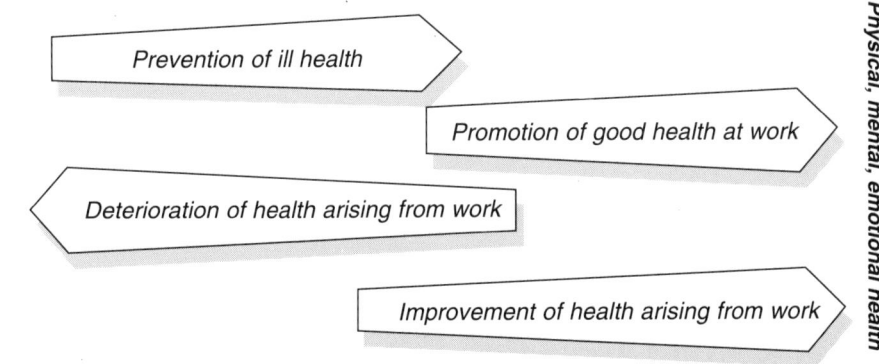

Losses expressed by:
- *ill health and injuries*
- *damage to property, plant, products and the environment*
- *loss to process and increased liabilities*

Positive benefits expressed by
- *reduced absenteeism*
- *improved job satisfaction*
- *general health and efficiency arising from increased commitment co-operation and competence*

Figure 2.3   Spectrum of occupational safety and health (source: HSE, 1991).

- In many cases workplace assessments are used to get an indication of safety and health risks at workplaces. This is the first step in a systematic reduction of risks, leading to healthier and safer workplaces.
- In the process of workplace assessment, the involvement of workers is a prerequisite. If well organized, workplace assessments offer good opportunities for worker participation.
- Good work contributes to the quality of life. Workplace assessments are an important step in the improvement process.

Some benefits will be visible in the short term. However, others may only become apparent after a number of years. Some occupational diseases take 10 years or more to develop. A good workplace assessment policy focuses on both short term and long term effects.

Basics and Fundamentals 3

3.1 Introduction

Workplace assessment has many variations. It covers a wide range of practices which may differ in several aspects, such as the methods used, the goals pursued and the topics that are covered.
There are some reasons why different views and practices have emerged. First, legislation differs among countries. Social security systems and the way occupational diseases are dealt with are quite different from one country to another. Workplace assessment must sometimes be adjusted to the specific requirements in a country.
Second, companies may have developed their own methods, tailor-made for their own problems. These assessments are related to the design processes and safety and health management within that company.
Thirdly, assessments can have different goals and scopes. Investigating noise complaints, for example, calls for a different type of assessment than supporting a design process in a greenfield site.
In this chapter the basic ideas behind workplace assessment will be made clear. Attention will be paid to the various ways in which workplace

assessment is developed and put into practice. This chapter will also serve as a guide to choosing the appropriate approach.

3.2 What is workplace assessment?

In essence, workplace assessment is the activity that answers two questions:
1. Which situations or activities within the company (or work process) cause or may cause undesirable effects (such as illnesses or accidents)?
2. Must something be done about it, and, if so, what actions should be taken and what are the implications?

The key elements of workplace assessment are:
- it covers all aspects of work: the tasks or activities to be performed, the persons that do the work, operating procedures, work flow, the organization, job content, the workplace and the working environment;
- it is primarily directed at consequences or effects the work has on people such as illness, discomfort or accidents, as well as positive outcomes (work satisfaction, wellbeing at work, better results from work);
- it is an action-oriented process, of which the actual investigation of work and work situations is a part; other parts are the evaluation and discussion of results, selecting priorities and setting up action plans for improvement;
- the process and activities are appropriate to the nature of the work and it is valid for a reasonable period of time;
- the primary aim is to improve working conditions: to combat risks for safety and health and (as a result) to get better results from work (such as productivity and quality);
- workplace assessments should provide solutions for problems encountered;
- the process is not only technical, but is part of the social context within the company and is part of management practice;
- the process is carried out in a systematic way.

Workplace assessment is not an instrument that investigates work situations and directly generates solutions. Its aim is mainly that of starting and structuring an improvement process based on a dialogue on working conditions within the company. Occupational safety and health then becomes a management practice in which the whole company participates.

Difference between Workplace Assessment and Risk Assessment

A difference is made between workplace assessment (WPA) on the one hand and risk assessment (RA) on the other. Workplace assessment takes a wide view and is very much oriented towards improvements in the work situation. All aspects of work, such as physical and chemical environment, ergonomics, occupational safety, mental strain, as well as organizational aspects, are considered. Quantification is not always necessary. Organising the process of assessment is included in the concept itself.

Risk assessment on the other hand, has a specific meaning. The main difference is that the latter is mainly concerned with the appreciation of risks. The goal is to quantify risks in order to decide on safer alternatives. Often risk assessment is focused on a limited number of scenarios, such as

Workplace Assessment	Risk Assessment
Workplace assessment is a wide concept aimed at finding possible hazards and improving the work situation.	Risk requires a precise definition. Several definitions (depending on the context) can be found.
In many cases qualitative, but also quantitative when needed.	Aimed at quantification. Risks are calculated in order to give an indication of the acceptability of certain risks.
Involves a wide range of topics, some of which are of a qualitative or subjective nature. Covers safety and health risks as well as wellbeing at work.	In many cases focused on major hazards and technical safety risks. In some situations wider meaning (for instance in Framework Directive).
Basic workplace assessments require some basic knowledge or experience. For detailed assessments specialists may be needed.	Generally risk assessments should be done by specialists.
Positive outcomes of work (for the worker: job satisfaction, health; for the company: better performance) are addressed as well	Primarily focuses on negative outcomes.

Table 3.1 Differences between workplace assessment and risk assessment.

the explosion of a gas tank or an emission of toxic substances. Here, workplace assessment is seen as a company instrument in the first place, whereas risk assessment has an important external function as well.

3.3 National context

National administrative traditions and regulatory systems have great influence on both the form of the workplace assessment instruments and the ways in which workplace assessments are carried out. In particular, in those countries in which self-regulation of the employer and employees with respect to health and safety is an important objective, assessments will primarily be carried out by the employer in cooperation with the workers. Health and safety experts only deliver supplementary support (UK, Netherlands, Denmark).

Other remarkable examples of national factors are, for instance:
- the dualistic security legislation in France and Germany: importance of occupational diseases, employers' liability
- extensive technical standardisation practices in Germany (accident prevention regulations, DIN and TÜV-standards used as preventive goals);
- framework legislation in Holland puts emphasis on the global assessment and (ergonomic) design practice, with special attention paid to the ways tasks are carried out;
- UK: existing safety practices in high-risk industries (nuclear power stations, chemical industries);
- traditional differing roles of trade unions in various countries;
- the UK and Ireland: traditional distinction between identification of hazards and (the evaluation of) risks;
- UK, Ireland, Denmark: the importance of codes of practice instead of detailed legislation;
- the role and nature of control bodies varies from one EU country to another.
- In Italy the concept of risk assessment has a specific meaning. Once a year a meeting must be set up between the employer, the head of protective and preventive services, the occupational physician and the

> **Example 3.1 Occasions on which workplace assessments may be started**
>
> 1. There are complaints about working conditions.
> 2. The company wants to start preventive policies.
> 3. An investment or design project provides an opportunity.
> 4. An incident or accident has occurred.
> 5. The company wants to make a start with quality management.
> 6. There is a change in legal obligations.

workers' safety and health representatives, in order to discuss any issue related to risks at the workplace.

3.4 Goals

General goals
As indicated in chapter 2, workplace assessments can have multiple goals. Examples of some general goals:
- starting point for preventive policies;
- part of safety and health management;
- collection of information on work in design projects;
- a means towards workers' participation;
- enhancing commitment from workers;
- complying with legal regulations;
- assessment of new organizational schemas;
- investigation of liability problems.

In most cases goals relate to one another in a tree-like structure. Every goal is a means to attain other, more general, goals. For instance, workplace assessment may be carried out as part of a preventive policy. Prevention is needed to lower the number of accidents and the sick leave rate, which in turn contribute to lower costs. Lower costs are required to increase profits or to maintain a competitive position. At the same time, compliance with most legal regulations is achieved.

Specific goals
Workplace assessments can have goals that are specific for one situation. For instance when workers have complaints about their workplaces, an

assessment must indicate the severity, the cause and suggest possible ways to solve the problem.

3.5 Types of assessments

The form that is chosen for a workplace assessment depends on its goals and the stage at which an assessment is started. For the investigation of physical workload in heavy work and an assessment for redesign, for instance, different types of assessments should be used. In this section four types of assessments will be explained:
- inspection of work situations: general methods of finding working condition problems at workplaces;
- thematic assessments: methods to investigate particular aspects of working conditions;
- task-centred assessments: focuses on the way tasks are performed;
- strategic or organizational assessments: investigation of management practices concerning occupational safety and health.

Inspection of work situations
Many instruments (mostly checklists) have been developed in order to investigate workplaces with respect to working conditions. In these instruments hazards of all kinds are evaluated in a systematic manner.

An important advantage of checklists for workplace inspections is that these are easily handled by non-experts. It is possible to get an impression of possible hazards in only a short time and without thorough training. It is this quality that makes the inspection of work situations at a global level suitable for small and medium sized enterprises. Sector-specific checklists are useful for small companies, especially for providing a global assessment.
Most of the inspection checklists tend to focus on the workplace and the working environment. Less attention is paid to hazards that arise from certain activities or working practices. Indeed, taking tasks and activities into account would require some form of task analysis, which makes the instruments more complicated, less easy to handle and more the domain of the specialist.
A common way of working is the 'cascade approach'. Simple and global instruments are used to get a global assessment. Where situations are found which may be hazardous or an assessment requires the use of more specific

methods, a more extensive (thematic) assessment is carried out. An example of this approach is illustrated in figure 3.1. Parts of the actual checklists are given in figures 3.2 and 3.3.

Task-centred assessment
In contrast to inspection techniques, task-centred assessments can be used. Whereas inspection methods focus primarily on the work situation and the environment, task-centred assessments start with an analysis of the work and the activities itself. The aim is to collect information on the 'real work', the way tasks are performed by workers in everyday practice. Often, it turns out that the actual way of working differs from the way which was envisaged by technical designers or management. Sometimes workers change procedures or develop practices that are quite different from job specifications.

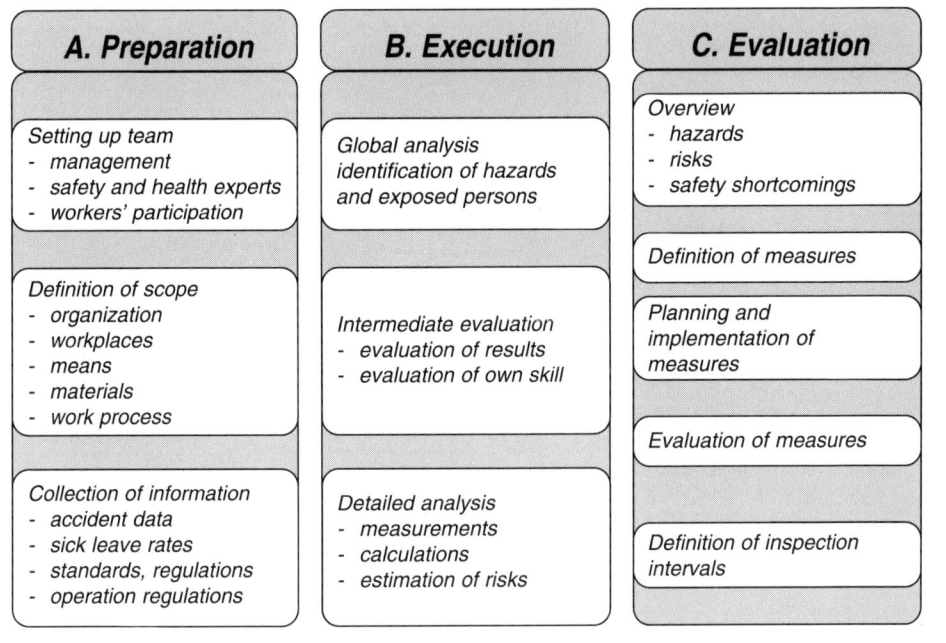

A. Preparation	B. Execution	C. Evaluation
Setting up team - management - safety and health experts - workers' participation	Global analysis identification of hazards and exposed persons	Overview - hazards - risks - safety shortcomings
		Definition of measures
Definition of scope - organization - workplaces - means - materials - work process	Intermediate evaluation - evaluation of results - evaluation of own skill	Planning and implementation of measures
		Evaluation of measures
Collection of information - accident data - sick leave rates - standards, regulations - operation regulations	Detailed analysis - measurements - calculations - estimation of risks	Definition of inspection intervals

Figure 3.1 Inspection of working situations according to the cascade approach: Grundmethodik Gefährdungsanalyse, general overview of the instrument and visualisation of the structure. (Adapted from: Bundesanstalt für Arbeitsschutz, 1993)

Phase	Activities	Contents	Instruments
preparation	Setting up assessment team	management, safety and health experts, workers, medical officer	
	Definition of scope	organization, departments, production lines, workplaces, machines	
	Collection of information	safety management system	Checklist
		relevant company regulations	
		sector related hazards and safety measures	
		company data (accidents, sick leave, occupational diseases)	
		survey among workers	Questionnaire
execution	hazard identification - global analysis - systematic analysis - safety behaviour	- sorts of hazards - sources of hazards - exposed persons	- list of hazards types - list of hazard characteristics - checklists
	evaluation of hazards - obligations - possible improvements - required improvements - priority, necessity	criteria: - legal requirements - state of technical know-how - guidelines - exposure time and frequency - possible consequences	schematic form
	if needed: special assessments	- work process analysis - task analysis - measurements - calculations	
elaboration	definition of measures for improvement	technical, organizational measures	
		time planning, deadlines	schematic form
		responsible persons	
	report		schematic form
realisation	implementation of measures		
	review of efficacy	implementation	
		effects	
		inspection intervals	

Figure 3.2 Inspection of working situations according to the cascade approach: Grundmethodik Gefährdungsanalyse, global checklist. The shaded areas illustrate the sequence of an assessment process. The sequence is continued in figure 3.3. (Source: Bundesanstalt für Arbeitsschutz, 1993)

Hazard characteristics

hazard category	hazard, hazard source	hazard characteristics	checklist nr.
1 mechanical hazards	1.1 unprotected moving parts	- pinching - shearing points -	1.1
	1.2 dangerous surface	- sharp edges - rough surfaces -	1.2
	1.3 transportation	- collisions -	1.3
	1.4		1.4
2 electrical hazards	2.1 touching live parts 2.2 touching parts that accidentally carry high voltages 2.4 very high voltages 2.5 interference arcs 2.5 static electricity	- inadequate isolation - failing protective measures - too low safety distances - thermal effects - ... - friction or flowing liquids	2.
3 biological agents	- ingestion of dangerous biological agents - skin contact - infections, allergic reactions, poisonous effects	- working with biological agents (viruses, bacteria, genetic material)	3.
4. ...	.....	......	....

Detailed checklist 1.1: unprotected moving parts

nr.	Questions	Criteria	Regulations
1.	Can technical installations give rise to hazards from moving parts?	- harmful effects of mechanical energy; - (sharp) edges; - inadequate safety distances	EN 294
2.	Are moving parts adequately protected?	- inadequate safety distances - poor maintainability -	
3.	Can hazardous situations emerge from normal situations?	- off normal situations: maintenance, start up, shut down, failures	
4.	Are adequate measures taken when presence near moving parts is required?	- presence - inadequate evacuation possibilities	
5.	Can hazardous situations easily be recognized?	- environmental factors (noise, lighting) - failing or inadequate alarms - failing indications - inadequate training or skills	DIN 4844 DIN 5035

Figure 3.3 Inspection of working situations according to the cascade approach: Grundmethodik Gefährdungsanalyse, detailed structure. The shaded parts of the table illustrate the way a hazard is investigated. Specific checklists can be used if hazards are present. (Source: Bundesanstalt für Arbeitsschutz, 1993)

All too often the workers are the adjustment variable to the errors and delays in the design and construction process. Mostly the operator adapts himself/herself to the system by changing the way of working and accepting inadequate working conditions. As a consequence, real work is different from prescribed work.

Task-centred assessments give a good view of the characteristics of work. It is made clear what workers do and what problems (in terms of working conditions) they experience. For instance a noise level of 75 dB(A) may not lead to hearing loss, but it can mask auditory signals or disrupt concentration (which in turn may give rise to safety or stress risks). In order to assess the hazards of the noise level, one has to take the task into account.

Task-centred assessments are especially useful in design projects or in situations in which human activity is crucial. Figure 3.4 shows the

Instrument	Short description
Grundmethodik Gefahrdungsanalyse (BAU, Germany)	Extensive instrument covering all aspects of working conditions. Consists of a planning model, overview of topics, checklists and references to standards and legislation (for illustration see figure 3.1). (Schütz et al., 1985)
Inspection Method working conditions (TNO & Ministry of Social Affairs, The Netherlands)	Checklists on 14 topics of working conditions (including stress and shift work), each of which has three levels: (a) checklists, (b) overview of possible measures and (c) guidelines, standards and legislation. (Ellens & Beumer, 1992)
Inspection Plus Package (NIA, The Netherlands)	Checklists to be filled out by the company on workplace and environment, means, people and organization, prevention and repression. (Stevens, 1993)
Evaluation of working conditions (INSHT, Spain)	Checklists for all aspects of work. Answers have three categories: sufficient, deficiency, major deficiency. In the checklists measures are indicated (see figure 3.8). (Arenaz Erburu et al., 1994)
Occupational risks, evaluation guide (Ministère du Travail, France)	15 checklists on working environment, workplace and safety. Each checklist has a theoretical introduction and reference to legislation and standards. (Ministère du Travail et de la formation professionnelle, 1994)

Table 3.2   Overview of some workplace assessment methods that include inspection of work situations.

importance of a procedure for a design project in which a task-centred assessment has a function.

Thematic assessments
Sometimes it is necessary to investigate an aspect of working conditions such as noise, physical workload or stress in detail. These specific investigations are called thematic assessments. There is a vast number of thematic assessment methods for every aspect of occupational safety and health. Some of the methods are rather complicated and must be done by experts. Measurements can be part of the assessment methods. An overview of some instruments is presented in table 3.3. Often this kind of assessment requires specific expertise. Nevertheless, on several topics simple instruments which can be used with little training are available.

DIAGNOSIS

of work in 'reference situations'

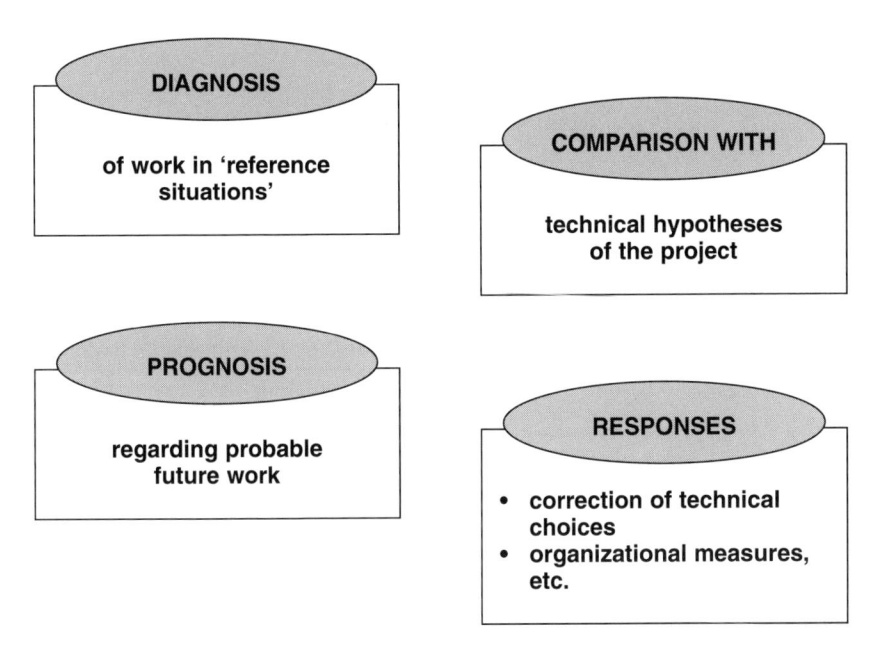

COMPARISON WITH

technical hypotheses of the project

PROGNOSIS

regarding probable future work

RESPONSES

- correction of technical choices
- organizational measures, etc.

Figure 3.4   Importance of knowledge about real work in reference situations. (Source: Du Roy, 1990)

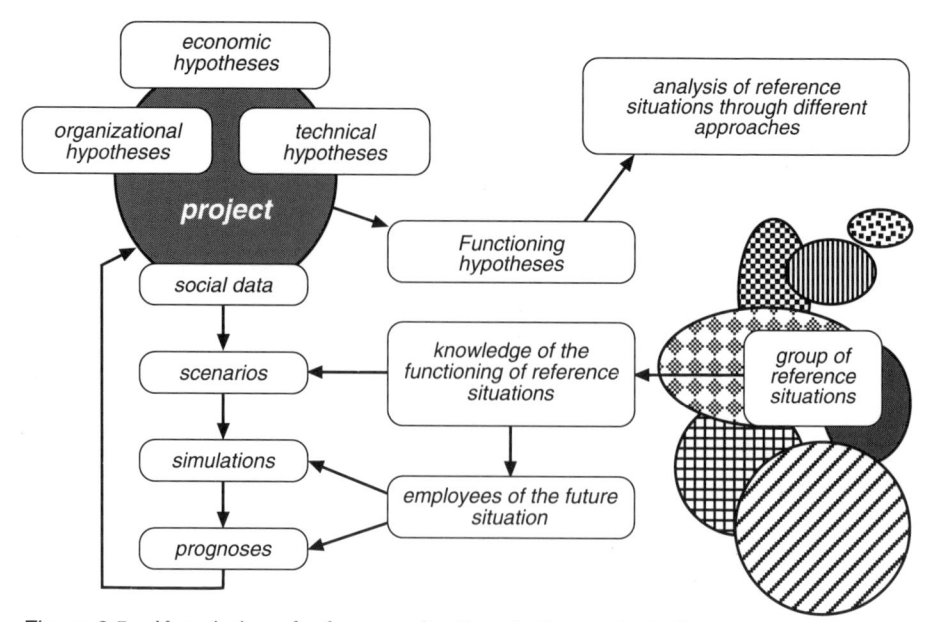

Figure 3.5   Knowledge of reference situations in the context of a design project (Source: Maline, 1994).

Topic	Examples of assessments
Noise	Protection of Workers against noise (Gamba & Abisou, 1992), standardized measurements methods
Climate	standardized measurements methods
Lighting	standardized measurements methods
Substances, gases, dust	standardized measurements methods
Stress	Job content checklist (Kompier & Levi, 1993)
Job content	WEBA, wellbeing at work (Pot et al., 1992)
Lifting, manual materials handling	Guide for manual materials handling, NIOSH
Computer workplaces	FNV Questionnaire for administrative work at computer screens (Berndsen & Vaas, 1991)
Computer software	The Software Checker (TCO, 1990)

Table 3.3   Some instruments for thematic assessments.

```
                    ┌─────────────────┐
                    │   decision to   │
                    │   take action   │
                    └─────────────────┘
           ↓                  ↓                    ↓
┌──────────────────┐ ┌──────────────────┐ ┌──────────────────┐
│ corrective action,│ │ preventive action,│ │opportunity of a  │
│   complaints of   │ │ legal obligations │ │  design project  │
│      workers      │ │                   │ │                  │
├──────────────────┤ ├──────────────────┤ ├──────────────────┤
│ problem analysis │ │ systematic assess-│ │ analysis of      │
│ - clarify complaints│ │ ment of all work │ │ situation to be  │
│ - preliminary analysis│ │ situations within│ │ designed        │
│ - people involved│ │ the company       │ │                  │
│ - define planning│ │                   │ │                  │
│                  │ │                   │ │                  │
│ execution of     │ │ planning          │ │ definition of    │
│ assessment       │ │                   │ │ reference        │
│                  │ │                   │ │ situations       │
└──────────────────┘ └──────────────────┘ └──────────────────┘
```

═══

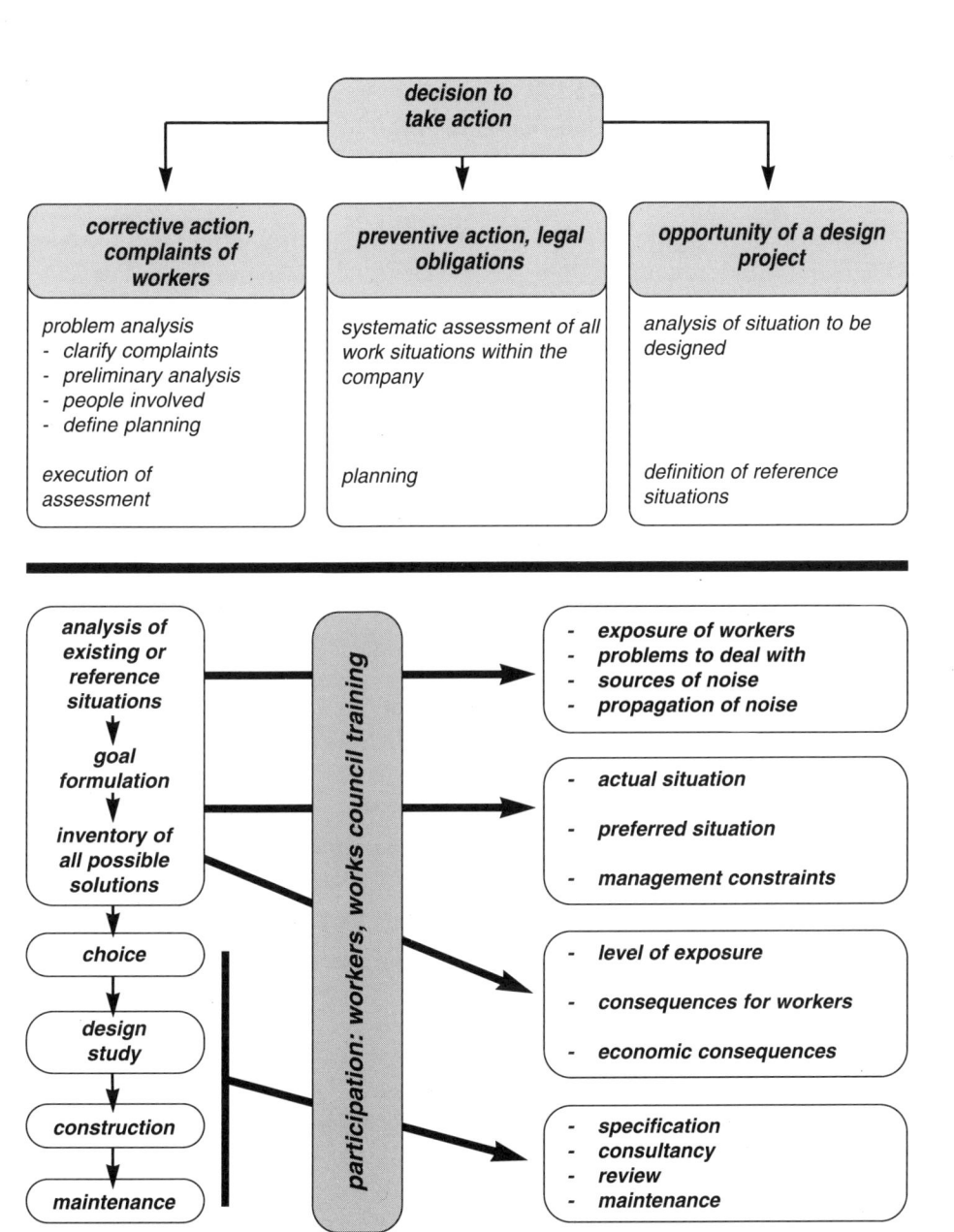

```
┌──────────────┐                        - exposure of workers
│ analysis of  │                        - problems to deal with
│ existing or  │────→                   - sources of noise
│ reference    │                        - propagation of noise
│ situations   │
└──────┬───────┘
       ↓
┌──────────────┐         p
│    goal      │         a
│  formulation │         r              - actual situation
└──────┬───────┘    ────→t
       ↓                 i              - preferred situation
┌──────────────┐         c
│ inventory of │         i              - management constraints
│ all possible │         p
│  solutions   │         a
└──────┬───────┘         t
       ↓                 i
┌──────────────┐         o              - level of exposure
│    choice    │         n
└──────┬───────┘    ────→:              - consequences for workers
       ↓                 w
┌──────────────┐         o              - economic consequences
│   design     │         r
│    study     │         k
└──────┬───────┘         e
       ↓                 r
┌──────────────┐         s              - specification
│ construction │    ,    ────→          - consultancy
└──────┬───────┘         w              - review
       ↓                 o              - maintenance
┌──────────────┐         r
│ maintenance  │         k
└──────────────┘         s
                         
                         c
                         o
                         u
                         n
                         c
                         i
                         l
                         
                         t
                         r
                         a
                         i
                         n
                         i
                         n
                         g
```

Figure 3.6 Model for noise assessment, starting from three different situations. The upper part of the diagram (above the line) gives the 'WHY', the lower part illustrates the 'HOW' of a thematic assessment.
(Source: Gamba & Abisou, 1992)

JOB CONTENT CHECKLIST

This checklist contains 19 questions which can be answered by every department. With a little editorial adaptation the questions can also be applied to every position.
Fill in 'yes' if you agree with the question. Subsequently, a total score can be calculated by adding up the 'yes' scores. The total score must be as low as possible. The higher the number of 'yes' answers, the greater the number of problems in the area of job content and the organization of work. Every 'yes' answer deserves separate attention.

		yes	no
1.	Short cyclical work is common. A task is short and cyclical if the same task repeatedly has to be started within 1.5 minutes of the last task; that is when the cycle is shorter than one and a half minutes.	☐	☐
2.	Dull or monotonous tasks are common (tasks which quickly become routine).	☐	☐
3.	Tasks which require intense concentration (from which one cannot remove attention) are common.	☐	☐
4.	The work in the department is strongly segmented. Every individual makes a small contribution to the 'product' of the department.	☐	☐
5.	The work is emotionally demanding, for example because of social contacts with patients, customers, pupils and the like.	☐	☐
6.	Work frequently takes place under pressure of time, deadlines have to be met, or production standards are difficult to achieve.	☐	☐
7.	There are lonely jobs in the department.	☐	☐
8.	It frequently occurs that the organization of the work and/or the work schedules is incorrect.	☐	☐
9.	.....	☐	☐
...	.....	☐	☐

Figure 3.7   Assessment of stress problems, part of a checklist (Source: Kompier and Levi, 1993).

Strategic or organizational assessments
In order to get a deeper insight into the working conditions within a company it is useful to understand why conditions are the way they are. For instance, poor lighting conditions may be caused by lack of maintenance, unhealthy working practices may have been caused by insufficient training.

Strategic assessments investigate the organizational context of working conditions. The goal is to analyze organizational procedures that influence working conditions. The following subjects may be included:
- operational management, the organization of work;
- assessments or inspections;
- maintenance;
- purchase;
- design and construction;
- knowledge;
- company culture;
- training.

Strategic assessments are in some way similar to audit instruments (see chapter 6.5). The main difference is that strategic assessments can be part of a global assessment, are more or less informal and can be carried out by the company itself. Strictly speaking, audits are more formal, and are carried out by an external expert. Audits also investigate whether procedures are carried out properly.

Choice of assessment type
The choice for a assessment type depends on several factors, for instance:
- the purpose of the assessment, whether there are any secondary goals;
- the context:
    = investigation of complaints;
    = part of a design project;
    = starting point for a safety and health management cycle;
    = compliance with legislation;
- the state of knowledge and experience within the company;
- the organizational structure preferred.

In some situations a clear distinction is made between the identification of hazards and the evaluation of risks. This distinction can be useful, in particular if the acceptability of certain risks needs discussion. Instruments which are used to inspect work situations often make implicit evaluations of risks.

PREVENTIVE MANAGEMENT			
1. RESPONSIBILITIES, FUNCTIONS			page [ ][ ][ ]
Date [ ][ ][ ]		Date next revision [ ][ ][ ]	
Completed by			

1. Management of the firm has made known its concern about the prevention of occupational risks and wishes to facilitate suitable means to improve working conditions	yes	no	The actual prevention of occupational risks starts with the consciousness of management and communication to the entire staff
1.1 The statement has been communicated to the entire staff of the company	yes	no	The actual prevention of occupational risks starts with the consciousness of management and communication to the entire staff
2. The principles for the development of the policy of prevention of occupational risks and the improvement of working conditions are written down	yes	no	The commitment for occupational risks on part of management has to be communicated in a statement of the principles of action
2.1 The statement has been communicated to the entire staff of the company	yes	no	The commitment of occupational risks on part of management has to be communicated in a statement of the principles of action
3. The tasks related to prevention of occupational risks corresponding to every level of the organization are clearly defined	yes	no	All issues have to be translated in action. It is necessary that the whole organization takes care of adequate working conditions in every workplace
3.1 The tasks concern the entire staff	yes	no	All issues have to be translated in action. It is necessary that the whole organization takes care of adequate working conditions in every workplace
4. The responsibilities concerning prevention of occupational risks are clearly established for all organizational levels	yes	no	It is necessary that the responsibilities concerning prevention of occupational risks are established, known and compulsory
4.1 The responsibilities concern the entire staff	yes	no	It is necessary that the responsibilities concerning prevention of occupational risks are established, known and compulsory
4.2 There are demands for control of those responsibilities for the entire staff	yes	no	It is necessary that the responsibilities concerning prevention of occupational risks are established, known and compulsory
5. The management of the company shows - with its daily behaviour - commitment for working conditions of all personnel	yes	no	It is important to show concern with facts, for instance by inspecting workplaces, investigating the accidents or participating in meetings

Figure 3.8 Strategic assessments. Part of the method for the evaluation of working conditions. (Source: Arenaz Erburu et al., 1994)

Situation	Appropriate assessment type
The company wants to have an early indication of problems on working conditions	inspection of work situations
The company is working on new production facilities	task-centred assessments
A design process is started	task-centred assessments
There are specific complaints about some aspects of working conditions (e.g. noise, climate or physical workload)	thematic assessments; choice of method according to the problem
There are vague complaints about working conditions	inspection of work situations; thematic assessments if probable hazards are identified
The company wants to investigate reasons for high sick leave rates	inspection of work situations; thematic assessments if probable hazards are identified
An incident has occurred and the company wants to take preventive measures	thematic assessments; task-centred assessments
The company wants to improve performance on occupational safety and health management	strategic or organizational assessments
The company wants to improve working conditions in order to enhance quality, productivity or flexibility	first start: inspection of work situations; if hazards are identified: thematic assessments; to support design of improvements: task centred assessments; to ensure efficacy: strategic or organizational assessments
The company wants to comply with legal obligations	inspection of work situations; strategic or organizational assessments

Table 3.4   Selection aid for assessment types.

Getting Started with Workplace Assessment 4

4.1 Introduction

Getting started with workplace assessment involves organizing, planning and communication with stakeholders within the company.
Setting up workplace assessment must be seen as a process. Initially, attention and activities may be limited and restricted in scope. Eventually full safety and health management systems may emerge, which are supported by all members of the organization.

It is wrong to think that just applying any assessment method or having an outside expert doing an assessment will give satisfactory results. A good plan, which is supported by all stakeholders in the beginning, will improve the quality and can avoid delays, discussions and conflicts at a later stage.

When getting started the following aspects have to be organized:
- objectives, desired results (see section 4.2);
- staffing (section 4.3);

- participation of workers and workers' representatives (section 4.4);
- scope and subjects of the assessments (section 4.5);
- setting up criteria (section 4.6);
- planning (section 4.7);
- follow up (section 4.8).

This chapter discusses the aspects that have to be dealt with in setting up workplace assessments. By means of reference, a simple structure for workplace assessments is shown in figure 4.1. This structure can be used as a blueprint for the workplace assessment process.

There is no single best way to establish a workplace assessment programme in a company. Experience shows that externally imposed programmes or systems tend not to work. It is not useful to impose organizational structures that are unfamiliar to the company. Activities must be worked out within the company. As a result, each company should set up a strategy that is compatible with its way of working, and fits into the company culture. The outline presented in figure 4.1 can serve as a framework that can be adapted to the company's own situation.
But whatever the culture, a clear strategy of dealing with assessments is needed. In companies where methods for quality control exist, a similar way of working can be adopted for workplace assessment. Figure 4.1 depicts a structure for workplace assessment that is comparable to quality control systems.

4.2 Selecting goals and specifying outcomes

Before starting, the company should have in mind what should be achieved with a workplace assessment. It is important that the company has some form of policy. It is most effective to link the assessment policy to other issues such as quality, human resources management, environment and technological change. When the assessment policy is linked to other issues, workplace assessment has a purpose which serves overall company policies. Without a distinct purpose and motive, the assessments will not lead to useful results and energy may be wasted. More seriously, a negative experience may hamper future initiatives. Possible goals are 'getting information on hazardous workplaces, the number of employees that have

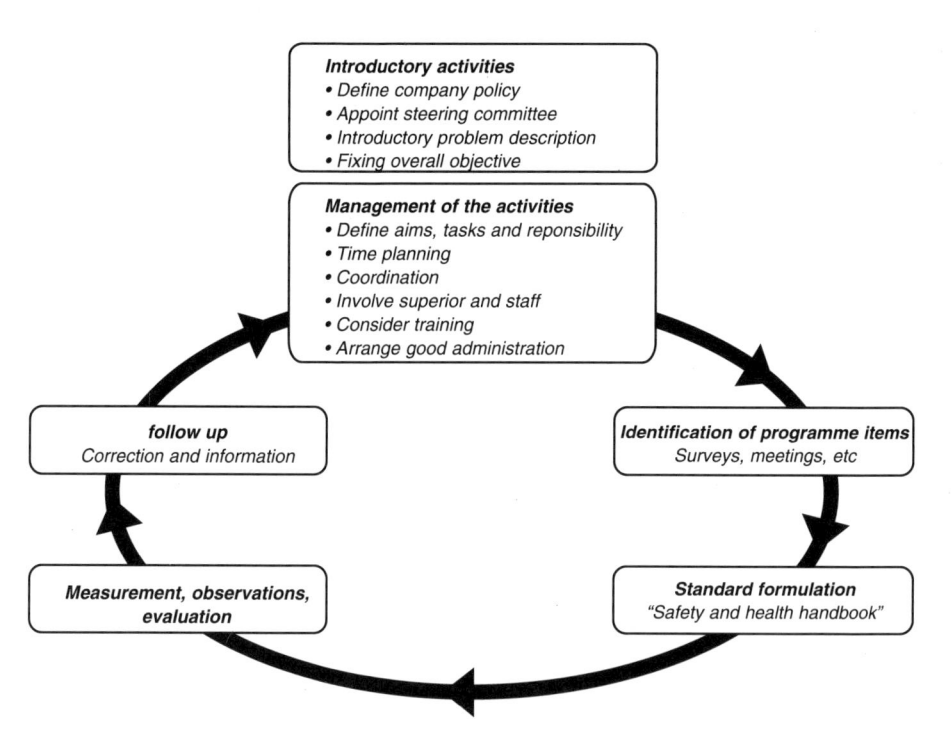

```
        ┌─────────────────────────────────────┐
        │ Introductory activities             │
        │ • Define company policy             │
        │ • Appoint steering committee        │
        │ • Introductory problem description  │
        │ • Fixing overall objective          │
        └─────────────────────────────────────┘
        ┌─────────────────────────────────────┐
        │ Management of the activities        │
        │ • Define aims, tasks and reponsibility │
        │ • Time planning                     │
        │ • Coordination                      │
        │ • Involve superior and staff        │
        │ • Consider training                 │
        │ • Arrange good administration       │
        └─────────────────────────────────────┘
```

┌─────────────────────────┐   ┌──────────────────────────────┐
│ follow up   │   │ Identification of programme items │
│ Correction and information │   │ Surveys, meetings, etc   │
└─────────────────────────┘   └──────────────────────────────┘

┌─────────────────────────┐   ┌──────────────────────────────┐
│ Measurement, observations, │   │ Standard formulation   │
│ evaluation   │   │ "Safety and health handbook" │
└─────────────────────────┘   └──────────────────────────────┘

*Figure 4.1   Schematic outline of a workplace assessment process.
(Source: Øland, 1993)*

Example 4.1 Overview of activities to perform at the start of a workplace assessment process.

In order to get this process started, some actions must be taken :

1. ensure commitment from top management and other key personnel;

2. ensure that involved persons fulfil their obligations;

3. someone must be occupied with the planning of the development and organizational learning process and make links between aspects of company policies;

4. the start of the development process must be clearly visible;

5. coordination of external support and cooperation with external bodies;

6. an organizational platform for the change process has to be established.

(Source: Jensen et al., 1993)

problems with workloads or collecting information in order to carry out a design project (see also section 3.4).

Setting goals is a participative activity. Official policy and the views of managers, workers and workers' representatives should be brought together. Often the views will diverge. One should be aware of this and it will be necessary to communicate or negotiate on this. Diverging aspirations in the beginning can easily lead to problems later in the workplace assessment process. When the goals are agreed upon, active support by management is needed.

The formulation of realistic goals is an important factor in the success of assessments. Both available budget and experience should be kept in mind when setting targets.

Carrying out an assessment may never be a goal in itself. Assessments are means by which to establish preventive policies and design improvements of working conditions. A real danger is that investigations may be used as an excuse to postpone or put off the redesign of workplaces. Commitment to implementing improvements is essential. Budget allocations for design or improvements and their subsequent implementation are necessary. This prevents the improvement process from stopping after the first workplace assessment has been carried out. Furthermore, budget allocations are a signal that working conditions are being taken seriously.

4.3   Persons involved

The workplace assessment process cannot work without well-defined tasks and giving responsibilities to several people. A large number of persons can be involved in workplace assessments. It is important to recognize that working conditions are in the first place the responsibility of (line) management. Management is the so-called 'problem owner' and has to see that workplace assessment is adequately carried out. The actual performing may be done and supported by others, preferably in cooperation with workers and lower management.

In an ideal situation all members and all departments of the organization have some function in workplace assessment. According to the principle of internal control, activities concerning working conditions should be part of everyone's daily activities (see figure 4.2).
Experts from inside or outside the company can support the assessment in various ways. An overview of possible participants and their activities is listed in table 4.1.

Like any other activity in a company, best results are obtained from people with high motivation. Often, activities for working conditions are performed by one person or by a small enthusiastic and committed group. It is of great importance to make use of this. Dampening their enthusiasm may set back all initiatives. Implementing new strategies will give the best results if these strategies fit current practices.

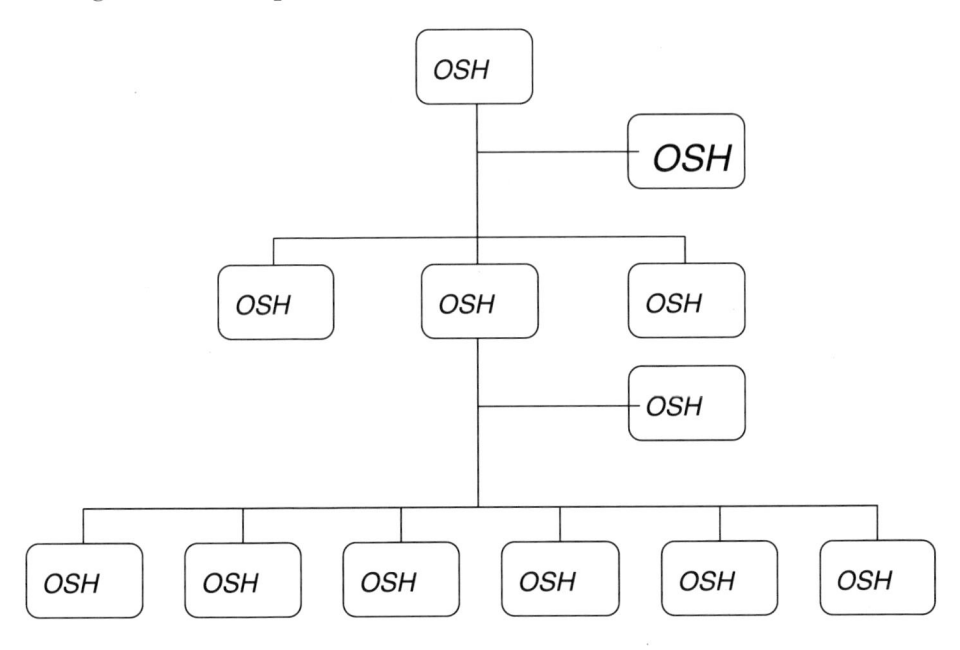

Figure 4.2 *Organization of activities for workplace assessment according to the 'internal control' principle. Specialists have a supportive role. Every department (including staff departments) has a responsibility in occupational safety and health (OSH).*

Persons or groups	Tasks and responsibilities
management and policy makers	- pursuing health and safety objectives and ensuring safe and healthy working conditions - devising health and safety policy - establishing strategies to implement policy - making a statement of commitment - proposing structure for planning, measuring, reviewing and auditing
middle and lower management	- carrying out (part) of assessment - stimulating activity of other persons involved - carrying out all necessary activities in order to take daily responsibility
workers' representatives	- advising on health and safety policy - taking part in policy formulation - taking part in discussions on the results of assessments
workers	- collecting and forwarding information - reporting on results - taking part in policy formulation - taking part in discussions on the results of assessments
experts	- advising on policy formulation - promoting safety culture - implementing assessments, monitoring results - performing and supporting the assessment - reporting on results - reviewing performance - training workers and managers
advisory bodies - suppliers - trade unions - insurance - consultants	- advising - providing information - training
authorities	- setting legal context - providing information on legal obligations - enforcing legal requirements - audit
sector organizations	- providing information - collecting data on a branch level - acting as advisory bodies

Table 4.1 Overview of possible persons or groups involved, tasks and responsibilities.

Role of experts
The role of experts in workplace assessments deserves some special attention. In some countries (e.g. The Netherlands) companies have to consult safety and health experts. It is easy for companies to farm out all safety and health activities to external services or specialized departments within the company. There is a real danger in leaving much of the safety and health activities to experts:
- there is no or little influence on important matters;
- only topics of marginal importance are left to specialized departments;
- managers and workers themselves develop little knowledge and experience and become dependent on experts;
- managers and workers have difficulty in formulating their own point of view and in speaking the technical language of experts.

As in good consultancy practice, safety and health experts must be supportive. They must help the company by doing their own workplace assessments as much as possible, for instance by providing information or training.
Some assessments, however, require specialist skills and must be done by experts. The task of the experts, then, is to communicate in such a way that workers and managers can form their own opinions on the results and conclusions.

4.4 Workers' participation

It is useful to make a distinction between direct participation of workers, indirect participation via works councils or (formal) workers' representatives, and participation via trade unions.

In most companies workers' participation is not very developed. In an ideal situation workplace assessment is a fully participatory affair. There is a real and extensive involvement of workers in any activity concerning workplace assessment.
There are two levels of worker involvement (WHIN, 1994):
- strategic level: unions and works councils are involved in the policy definition, specifying goals, the choice of methods and the like;

- operational level: the workers at the shop floor are involved in the assessment itself and the outcome.

In practice, on the operational level, several forms of workers' participation are found:
- Assessments are initiated and carried out by the workforce. Assistance may be obtained from an expert.
- Assessments are carried out by an expert. Workers provide the necessary information. Involvement mostly has an ad hoc character.
- Representatives of workers are involved in the assessment.
- The process of workplace assessment is organized by close cooperation between management, workers and experts (who may do some parts of the assessment).
- Some issues can only be assessed with direct and intensive involvement of workers. In particular a matter such as stress at work cannot be examined without the cooperation of workers.

There is a difference between the work as it is described, and the work as it is actually performed and experienced by the workers. This difference stems from fluctuations of external elements, such as physical environment (noise, climate, lighting), the time of day, the product, the organization of work and so forth.

These fluctuations lead to changes in the physical and mental state of the workers. Workers adapt their behaviour to changing circumstances. It is for this reason that a workplace assessment must also take the adaptive behaviour of workers into account. The cooperation of workers is naturally indispensable at this point.

Moments at which participation is essential:
- identification of hazards of workplaces and activities;
- evaluation of exposure to adverse environmental conditions;
- generation of solutions;
- introduction of improvements;
- evaluation of the assessment and its results.

Advantages
Apart from some more or less theoretical reasons, workers' participation has some important advantages:
- it capitalizes on knowledge of processes, tasks, problems; many problems are experienced by the workers daily which may not be apparent from the outside.
- it gets commitment from workers;
- it enhances acceptance of results;
- it facilitates cooperation in design and intervention projects as a follow-up.

The quality of a workplace assessment is highly dependent on cooperation with workers.

The organization of worker participation
In principle, workers' participation is a prerequisite for effective workplace assessments. New European legislation obliges employers to involve workers actively in the actual and strategic health and safety policy in the company. However, the statutory responsibility for risk assessment lies with the employer. In case of harm or injury, employers are liable. In practice big differences do occur in workers' participation. In some countries (e.g. UK) the trade unions are traditionally the normal sparring partner of management. In Italy, France and Spain many of the workers participating are trade union members, but they do not represent the trade union. In other countries like Germany and The Netherlands, works councils have legal participation rights and therefore are strongly involved in health and safety policies in the company.

A good starting point for a workplace assessment is a presentation, organised by involved persons and line management. In this presentation management can show its intentions. In the meantime information can be given on the workplace assessment activities:
- the objectives;
- global planning;
- persons involved.

Next, actual participation can be organized, for instance in the form of workgroups.

Example 4.2 Workgroups as a means of participation.

Workgroups can provide an effective way of participation of workers. However, there are some conditions that have to be met.

composition of the workgroup
In the workgroup different persons take part:
- workers' representatives;
- working conditions specialists, ergonomists;
- management.

The size of the workgroup depends on the nature of the problem and the size of the company.

mission of the workgroup
The mission of the workgroup must be clear. In a workgroup activities are oriented towards:
- analysis and measurements;
- participation in generating solutions for problems;
- monitoring workplace assessment process;
- communication

means
The members of the workgroup must be capable of performing their tasks. They need:
- adequate training;
- time for special activities;
- secretarial support.

	Work group	Steering group
Function	- setting up design criteria for problems concerning working conditions - working out recommendations	- discussing proposed solutions - negotiating possible modifications for improvement of working conditions - proposing solutions
Composition	\multicolumn{2}{c}{management (2 persons appointed by name) workers' representatives (4 persons appointed by name) ergonomists (5 experts from outside the company)}	
Other participants	- persons of departments, involved in the actual problem - experts	- various representatives (8 persons appointed by name)
Meeting frequency	every Tuesday morning	every 4 to 6 weeks

Figure 4.3 Workgroups in redesign project at 'Le Monde' (Source: ANACT, 1989; Sutter et al., 1992)

The Chair
Questions (see figure)

1 Is the seat at least adjustable in height between 41 and 53 cms? yes ☐
   no ☐

2 Does the back of the chair have a lumbar support (bolster in the back yes ☐
   of the chair that fits into the hollow of your back)? no ☐

3 Can you modify the depth of your seat? This is the distance between the
   back and the front of the seat. The depth of the seat can be modified by yes ☐
   moving the back of the seat or the seat itself forwards or backwards. no ☐

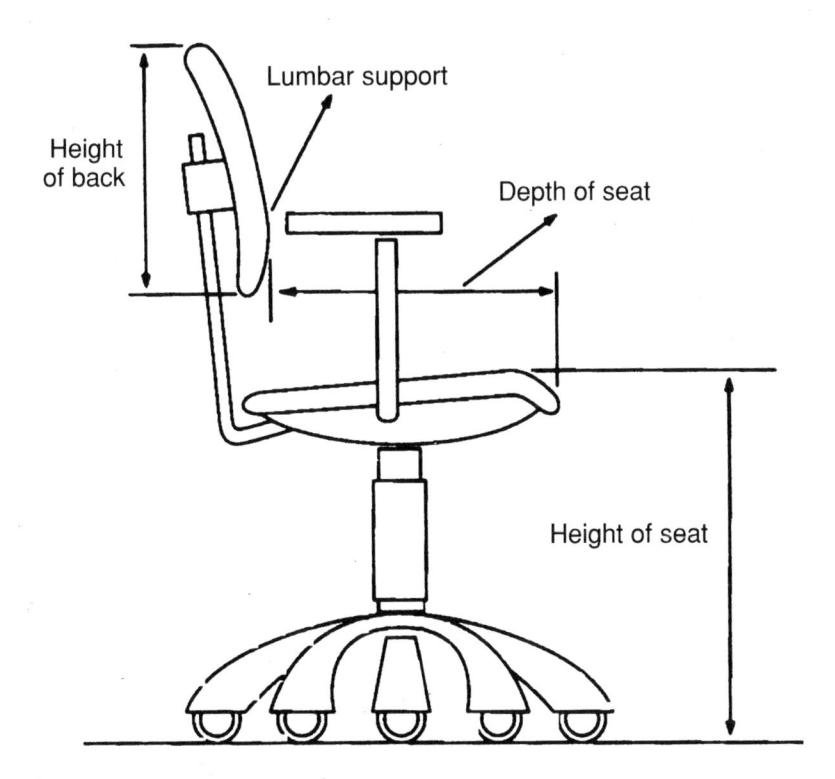

Figure 1: The office chair

Figure 4.4 Part of the 'do it yourself' checklist for the evaluation of administrative work at computer screens. (Source: Berndsen, Vaas, 1991)

Instrument	Described role of workers
Protection contre le bruit (Gamba & Abisou, 1992)	Extensive description of the way in which participation (in workgroups) can be organized
Einbeziehung Arbeitnehmer in die Gefährdungsanalyse (Lohrum, 1993)	Indications of the way in which participation can be organized at different phases of workplace assessment
Guidance on risk assessment at work (Kloppenburg, 1993)	Several notes on involvement of workers
Safety and Health Circles, Quality Circles (Ritter, 1993)	Role of workers in solving working condition problems

Table 4.2   Overview of practices and comments on worker participation.

Setting up workers' participation
Workers have some (formal) rights which affect how the involvement of workers is incorporated:
- Workers can gather information in an independent way. They may express an opinion of their own.
- Workers (or works councils) have the (formal) right to call in outside experts.
- Workers can influence the results and the way results are reported.
- Workers must have no fear for their positions and feel free to make statements about their work situation.

In practice, these (sometimes formal) rights can easily be incorporated in participative techniques.

Often the attitude of top and middle management is decisive for the success of worker involvement. A positive attitude and full cooperation is extremely important, as is shown by experiences with quality circles. The amount of formalization also plays a role. Quality circles and worker involvement in safety and health cannot be maximized if only formal procedures are applied.

A problem often encountered is that workers' participation is reduced to some "harmless" topics.

Simple practicable instruments can support or enhance worker involvement. For some topics simple 'check your own workplace' tools exist (see figure 4.4).

Cooperation of management and workers is extremely important. Experience with quality circles shows that the amount of formality is decisive.

Works councils
Because of potential conflicting goals and interests between management and workers, all safety and health issues in principle should be discussed in works councils or an equivalent at the firm level. This also has the advantage that health and safety at the shop floor level can be related to the more strategic firm level. The works council can play an important role in formulating the company's health and safety policy as well as in the regular evaluation of its efficacy.

Role of trade unions
In some EU countries, works councils or their equivalents don't exist (e.g. Great Britain, Ireland). In these countries trade unions traditionally have a strong position at the shop floor level (shop stewards, safety stewards). In those cases it is essential to give the trade union representatives a clear, albeit not exclusive, role as workers' representatives. Sometimes health and safety issues can be adequately negotiated in workplace bargaining.

In countries with well-developed works councils, the unions normally don't play an outspoken role at the shopfloor level, but at company level. In these cases it may be useful to give health and safety issues a place in company-wide or sectoral bargaining processes. For instance, in The Netherlands in a number of branches the making of workplace assessment instruments is discussed and further specified at this level. This is especially significant in the case of small and medium-sized businesses (for instance construction industry, farming and cleaning).

In sum: direct and indirect worker participation is a basic requirement for effective workplace assessments. It not only contributes to better solutions for health and safety problems, it also broadens their acceptability.

4.5 Selecting scope and subjects

Before starting a workplace assessment there may be a number of questions, such as:
- which part of the premises of production site are to be taken into account;
- which activities will be considered;
- which subjects or themes will be addressed.

Assessments should be suitable and sufficient. Putting in more effort than necessary would mean loss of time and inefficient use of resources.

The organizational effort (such as the size of workgroups and the detail of planning) should not be excessive.

A first and global identification of hazards may entail every workplace, all activities and persons and all aspects of work. For thematic assessments or detailed assessments the scope can be limited to some activities, topics and departments. Every aspect should be in proportion with the associated risks. There is no use in an extensive assessment on some topic if there are no complaints or serious consequences. A quick scan may be sufficient.

Possible criteria in defining the scope and the level of detail are, for instance:
- indications about the nature of hazards, gravity of possible consequences;
- number of persons exposed;

The scope of an assessment should be well defined (and communicated to all persons involved). False expectations can thus be prevented.

Unusual circumstances should be part of an assessment as well. Many accidents tend to happen during non-routine activities. Examples of these circumstances are:
- plant modifications, extensive maintenance;
- contractor activities;
- procedures which only occasionally take place (e.g. emergency procedures);
- breakdowns, unplanned events.

It goes without saying that workplace assessments take all workers (women and men) into account. Some groups of workers, however, deserve special

attention, because for these people the workplace may have special or aggravated risks:
- young and elderly workers;
- pregnant and nursing women;
- handicapped or (chronically) ill people;
- external workers (contractors) or workers on short contracts;
- foreign workers (that have special cultural backgrounds or have problems with the language).

As indicated, workplace assessment should cover all aspects of work. In some cases, however, it is useful to narrow the scope to some aspects that cause problems. Possible subjects for assessments:
- safety (mechanical, heights, electrical, fire and explosion, chemicals);
- physical workload (working postures, physiological load, movements and exertion of forces, manual material handling);
- ergonomics of workplaces and workstations (layout, sizes, controls, information display);
- working environment (noise, lighting, vibrations, climate, radiation);
- chemical substances, gases, vapours, dust;
- biological agents (micro organisms, germs, viruses, toxins);
- mental workload (cognitive and perceptive workload);
- job content (cycle times, learning opportunities);
- emotional strain (aggression, sexual intimidation);
- stress (control demands and control capabilities);
- organizational factors (work schedules, hierarchies, social support, management styles).

4.6 Setting up criteria

Specification of criteria for working conditions (e.g.: 'the noise level should not exceed 80 dB(A)') must be part of a well worked-out company policy on safety and health. Difficult discussions can be prevented if criteria are a clear and negotiated part of the working conditions policy.
It is possible to specify criteria with respect to a number of aspects of work beforehand. In some situations, however, setting up criteria may be difficult. For instance, when a company has little or no experience in formulating safety and health policies and possible hazards are unknown, it is more

Model	Short description
Protection contre le bruit (ANACT)	Model for planning noise assessments in which cooperation with workers is an important part (see figure 3.6)
Guidance on risk assessment at work (DG V)	Extensive model for planning workplace assessments (see figure 4.5)
Risk Assessment Toolkit (Loughborough University)	Planning model based on assessment by workgroups in which workers have a role
ABRIE (Ruigewaard, 1993)	Model for assessments used by occupational health services

Table 4.3 Overview of models for the planning of workplace assessment.

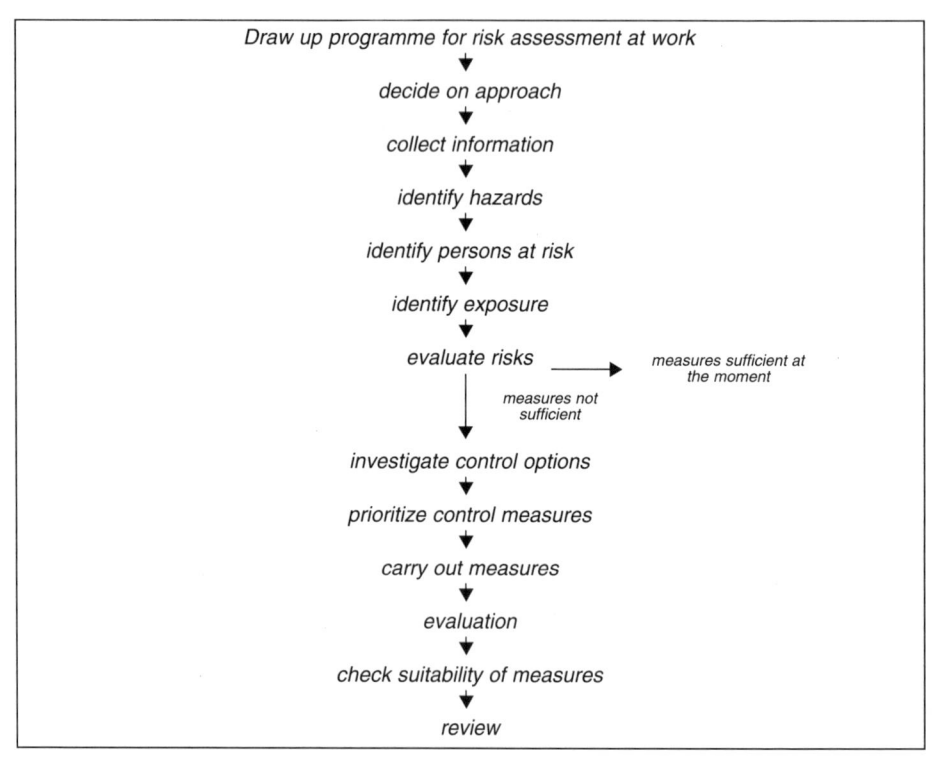

Draw up programme for risk assessment at work
↓
decide on approach
↓
collect information
↓
identify hazards
↓
identify persons at risk
↓
identify exposure
↓
evaluate risks ⟶ measures sufficient at the moment
↓ measures not sufficient
investigate control options
↓
prioritize control measures
↓
carry out measures
↓
evaluation
↓
check suitability of measures
↓
review

Figure 4.5 Possible structure of the workplace assessment process (Source DG-V, 1993).

fruitful to start a discussion about criteria after the assessment. Eventually a company handbook containing safety and health standards can be compiled.

With respect to criteria there some pitfalls that have to be avoided. Points that should be kept in mind are:
- criteria may be implicit in the method used,
- checklists are mostly based on legal requirements or standards, which may be out of date. A firm may wish to use other (more restrictive) standards;

Example 4.3   Eleven-steps approach to planning workplace assessments

Step 1
Nominate a risk assessment leader or coordinator who may brief senior management.

Step 2
Establish a risk assessment team. This logically requires organizations to ask 'Who have we already got and who is missing from the team'.

Step 3
Ensure that all team members are briefed and have appropriate training.

Step 4
Undertake an organizational analysis to produce a list of activities and employees/job titles. Extend the list to include a review of all 'non-employees' who may be affected by your activities. Consider the physical boundaries of the organization. Nominate specialists for each of the key areas of your activity (it may be useful to assign each area a number for ease of recording).

Step 5
Review all existing assessments and define the scope of future assessments and coordinate activity.

Step 6
Agree on the methodology for assessments and plan against agreed timetables.

Step 7
Collect and collate all relevant information and existing documentation.

Step 8
Estimate and evaluate risks and agree on action plan.

Step 9
Record assessments and collate information (and prepare any necessary documentation). Implement action plan and act on any priority areas immediately.

Step 10
Define and implement a monitoring system (audit and review) and agree on criteria for re-evaluation.

Step 11
Share information with all employees and anyone affected by your operations.

- for some topics, no standards or guidelines are available.

The use of criteria is also discussed in section 5.3.

4.7 Planning

In most companies some information on working conditions is available. In many cases there is some experience. It is essential that further workplace assessment activities make use of this experience. A first step should be the collection of previous experiences.

A timetable in which actions, responsible persons and deadlines are written down is of great help in keeping the process going.

4.8 Follow-up

Agreements on follow-up must prevent the problem-solving process ending after the completion of the workplace assessment. Planning a course of action is the next step.

Workplace assessment may never be an excuse to postpone or put off further action. In some cases it may be useful to set apart a budget for the implementation of improvements beforehand. All too often investigations are used as an excuse for taking no preventive or corrective action.

Carrying out Workplace Assessments 5

5.1 Introduction

The body of the process of workplace assessment consists of collecting information on hazards concerning work, the work process and the working environment and then determining whether some situations should be improved. In many situations a step-wise approach can be adopted. The hazard identification step mainly concerns gathering as much information as is needed (see section 5.2). The goal is to make an overview of situations or activities that may have adverse effects on safety, health and wellbeing at work. The next step is to evaluate the hazards identified and to determine whether hazards encountered can have serious consequences (section 5.3). Finally, a scheme for follow-up should be designed. The question of which problem to tackle is also important (sections 5.4 and 5.5).

5.2 Collecting information

There are many sources of information and many techniques for searching for hazards. Often, a lot of information is already available within the company, such as sick leave rates and accident rates. Some hazardous situations may be known already. Some information (noise data, chemical substances, and in some countries sick leave or health reports) must be available in the company by law.

It pays to find out what information can easily be obtained. Information can be found in, for instance, company reports, branch studies and company statistics on sick leave and accidents. More elaborate techniques for hazard identification (such as the use of checklists, questionnaires and measurements) can be used as a next step to get more details or to fill in gaps.

Some information sources use company data:
- observation;
- checklists;
- workgroups;
- measurements;
- questionnaires and interviews.

Other are directed at outside information:
- databases and statistics on accident and health;
- sector studies and literature.

It is preferable and often essential to make use of several techniques for information collection in the same assessment. For instance, an observation of the work process by an expert can support opinions gathered from a questionnaire survey. Measurements (e.g. for noise or climate) may be needed to obtain more details about complaints.

It is important to keep in mind the goals of workplace assessment. Efforts devoted to the collection of information should be proportionate with the goals, the size of the firm and the risks for safety, health and wellbeing (these can be reliably estimated in advance).

Observation of the work process

Observation of the work process, or job or task analysis, is the most thorough method of identifying hazards. Observation is time-consuming and requires some experience, but there some important advantages:
- observation is activity-oriented, i.e. the work situation is considered in

> **Example 5.1**
>
> **Workload assessment in the chemical industry.**
> In a company in the chemical industry a question was raised about workload. An assessment was started in order to find out if there was a workload problem.
> Workload assessment was done by use of multiple methods and data sources:
> -questionnaire about work, working conditions, shift work and health among all employees; questionnaire analysis was done by the consultant
> -analysis of administrative data of sick leave, turnover, incapacity for work and overtime (done by the consultant and the company;
> -analysis of working schedules (performed by the consultant);
> -inspection of work situations, supplemented by short interviews of workers (performed by the company in cooperation with the consultant).
>
> *(Source: TNO-PG practice)*
>
> **Workplace Assessment at DAF**
> The Dutch truck manufacturing company DAF combines inspection of workplaces with the opinions of workers. Inspection of workplaces is done by means of a checklist filled out by lower management and (if necessary) measurements. The opinions of workers are collected by means of a questionnaire. Part of the assessment is a comparison between the results of the inspection and the questionnaire. Differences between the two are subject to discussion.
>
> *(Source: Peters, 1993)*

relation to the activities performed;
- observation focuses on the real work, i.e. activities are observed as performed and not as prescribed.

The latter is particularly important. Workers may have altered their ways of working, giving rise to new and unforeseen hazards.

Observation of the work process must be carried out with the full cooperation of the workers. An introduction to the purpose of the observations, the way observations are performed and what is done with the results is essential. Some form of feedback from the observer to the workers is also required. The workers should know what is recorded and the analyst

Op.	STAGES OF ACTIVITIES OR DESCRIPTION OF THE OPERATIONS	Posture			Effort		
		lev	freq	dur	wt	freq	dur	lev
							Av	sum	sum	lev

Observations:

POSTURE

POSITION OF THE TRUNK		POSITION OF THE HANDS AND ARMS	OTHER POSITIONS	LEVELS (Duration or frequency)	
FRONTAL	LATERAL				
sitting vertical trunk	vertical without torsion	hands lower than the heart	dynamic posture	1	
standing vertical trunk	vertical without torsion	hands lower than the heart	standing position with minor displacements or alternating with seating position	2	
lightly bent less than 30	lightly bent sidewards or light torsion	hands between the heart and the shoulders	Static position	3	
bent between 30 and 45	bent sidewards or major torsion	hands at head level, or arms stretched out	climbing objects, important displacements	≥ 1/3 tcy ≥ 100 f/h 4	< 1/3 tcy < 100 f/h 3
bent more than 45 bent backwards	bent severely sideways or torsion of about 90	hands above the head	Particular movements with risks for the lumbar spine	≥ 20% tcy ≥ 50 f/h 5	< 20% tcy < 50 f/h 4

Figure 5.1 Part of an ergonomic observation sheet suited for tasks with a cycle time of less than 6 minutes. In the lower part of the figure the analysis of postures is shown. (Source: Renault).

should verify his observations.

Observations are structured and carried out in a systematic way. A number of techniques is available, for instance the job safety analysis (see figure 5.1).

Checklists

Checklists can be regarded as simple tools for the identification of hazards, which are often part of a more elaborate protocol for workplace assessment. Checklists are suited for systematic and global identification of possible problems, especially in situations when there is no knowledge about working conditions. As checklists often give global information, more elaborate methods like measurements could be used in addition (cascade approach, see section 3.5). Checklists can be symptom-oriented (see for example figure 5.4) or factor-oriented. The latter is aimed at identifying possible factors of risks (see figure 3.7).

It should be noted that symptom checklists give little insight into the underlying causes of problems. As a consequence the opportunity for making improvements is limited. Checklists merely draw attention to a hazard.

It should be noted that checklists are always a trade-off between having a simple tool on the one hand and having a complete list on the other hand.

There are two kinds of checklists: closed checklists and open checklists. A closed checklist offers the assessor a complete set of items which can easily be checked. This type of instrument is especially useful for non-expert or occasional users. For some instruments limited knowledge or experience is sufficient so workers or (lower and middle)management may very well use this type of checklist. An example is given in figure 5.2.

Open checklists are more flexible and leave room for extensions. Mostly only the topics are given, so it is merely a list of aspects that have to be dealt with. The assessor has to adjust the checklist to the specific situation and fill in the details. Using this type of checklist requires more knowledge and experience.

Sometimes it is useful to construct a special purpose checklist. An open checklist is the starting point. The result is a (closed) checklist that can be used by workers or line management.

Workgroups, safety and health circles

Workgroups or safety and health circles offer good opportunities of involving workers in collecting information about work, workplaces and activities. It is important to organize a process in which participants are stimulated to bring forward all kinds of information and problems with respect to working conditions. In the initial phase of a workplace assessment especially, every idea or piece of information is needed. Techniques like brainstorming or brainwriting are useful in the search for hazards.

PLACE (company, premises, workstation):		
Date:		
Questions	YES	NO

- Have the lighting conditions been subject to complaints
    at the workstations?
    at the workplaces?
	☐
	☐
		☐
		☐

- Do the workers complain about visual fatigue?	☐	☐

- Does this visual fatigue refer to
    a certain period?
    a certain task?
	☐
	☐
		☐
		☐

- Are light sources or their reflections visible in the centre of the visual field at the workstation?	☐	☐
- Are the workstations exposed to direct sunlight?	☐	☐
- Do the workers complain about headaches or backaches?	☐	☐
- Can the workers individually adjust the lighting at their workstation in order to perform their tasks?	☐	☐
- Can the workers easily see information required for performing their tasks?	☐	☐
.....		
.....		

Figure 5.2   Closed checklist in order to find lighting problems. The checklist is part of a more extensive checklist for working conditions. (Source: Ministère du Travail, de l'Emploi et de la Formation Professionelle (France), 1994).

Usually, workgroups need a moderator. Someone has to structure meetings, make notes, draw conclusions and report. Often this role can be performed by an occupational health and safety specialist. Figure 5.3 illustrates the structure of workgroups that collect information concerning hazards at the workplace and draw up improvements.

In order to be able to influence decision-making, workgroups must have some formal status and be embedded in an organizational structure.

Interviews and questionnaires

Workers can be seen as the ultimate experts on their own workplaces. A workplace assessment is not complete without the suggestions and opinions of workers. In large companies questionnaires can be a suitable way of learning about hazards. Safety and health problems can be traced to departments (if the number of workers per department is more than about 20 persons). In small companies or departments, anonymity of workers cannot be guaranteed.

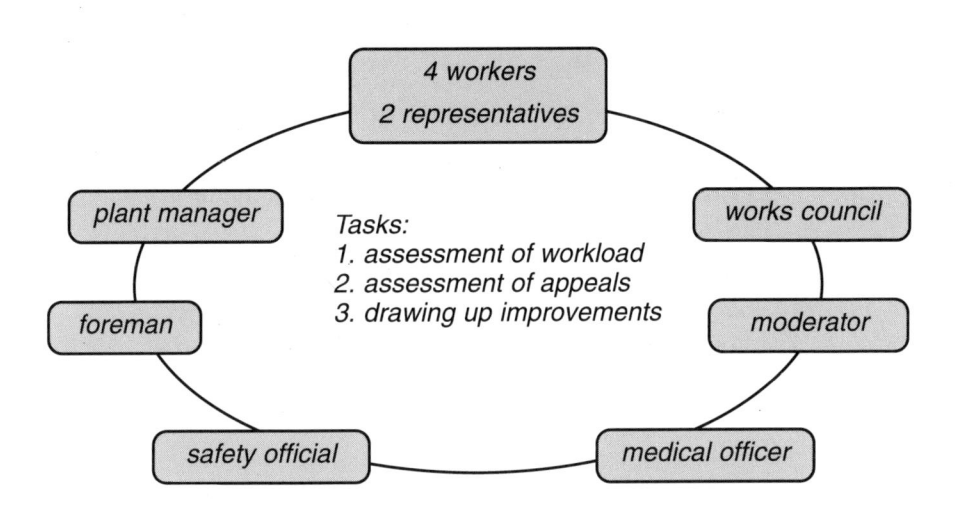

Figure 5.3 Safety and health circles at Asea Brown Boveri (Source: Huf, 1994).

> **Example 5.2   Health circles at a German steel mill.**
>
> Health circles have been tried out in a pilot project in a German steel mill. For every shift team a health circle was started. Six to eight meetings of about 1.5 hours in four weeks are sufficient. The meetings require high concentration, intensive preparations and intensive discussions. The results are mostly very practical proposals. It is important to present results at management level. Management support is indispensable.
>
> For each circle, about 20 strenuous work situations were identified and about 30 to 40 proposals for improvement were drawn up. Most of these were easy to implement, but sometimes there can be financial limitations.
>
> The costs of health circles and improvements are mostly low, between a few hundred and a few thousand D-Mark. The cost of one day's sick leave is about 600 D-Mark.
>
> *(Source: Ligteringen, 1994)*

Interviews provide more direct contact and leave more room for flexibility. In an open interview workers can be encouraged to identify possible safety and health hazards. A combination of observations and interviews is very useful. The assessor can ask for details about his observations, and feedback can be given to the workers. Interviews are, however, time-consuming.

Measurements

In order to obtain detailed information on working condition problems, measurements can be performed. Most types of measurements are complex and must be done by trained specialists. In workplace assessments, measurements are needed when a global approach indicates that there is a problem, and are also useful when doubts are expressed. Aspects of working conditions that are measured generally concern the physical and chemical work environment (e.g. noise, climate, toxic substances). However, measurements of mental and physical workload can also be done. Methods for the latter are not standardized, whereas most work environment measurements have been standardized.

Databases and statistics

There are different ways of using information from databases or statistics.

		almost always	often	some- times	hardly ever	never	relation with work	
							yes	no
1.	Do you have headaches?	☐	☐	☐	☐	☐	☐	☐
2.	Can you feel in your body when you get agitated about something?	☐	☐	☐	☐	☐	☐	☐
3.	Do you notice heart palpitations at low physical effort?	☐	☐	☐	☐	☐	☐	☐
4.	Do you have a feeling of fullness?	☐	☐	☐	☐	☐	☐	☐
5.	Are you tired quickly?	☐	☐	☐	☐	☐	☐	☐
6.	Do you notice dizziness sometimes?	☐	☐	☐	☐	☐	☐	☐
7.	Do you have a sensitive stomach?	☐	☐	☐	☐	☐	☐	☐
8.	Does little physical effort lead to lack of breath?	☐	☐	☐	☐	☐	☐	☐
9.	Do you have backaches?	☐	☐	☐	☐	☐	☐	☐
...	.....	☐	☐	☐	☐	☐	☐	☐

Figure 5.4 Part of the Volkswagen questionnaire on health and wellbeing. The complete questionnaire consists of 83 questions about health and lifestyle (Source: Volkswagen AG)

1. Data (generated by the company itself) can be used for reference within the company. Several sources can be useful:
   - statistics of sick leave and accidents can be used as a monitoring instrument, and improvements over time can easily be shown;
   - properties of dangerous materials which are used within the company;
   - accident investigation reports;
   - results of former workplace assessments.
2. Often, information that has been generated by companies is aggregated for branches or on a national level. This information is fed back to the companies. This can useful in the following ways:

- benchmarking, where the company can compare its own situation to what is usual in comparable companies;
- similar problems within comparable companies;
- data exchange between companies.
3. Occupational safety and health data on a wide variety of topics are collected by institutes, and a number of databases is available. Some examples of databases are given in table 5.1.

A complete overview of databases is given by the HASTE system (European Foundation, 1993). See figure 5.5.

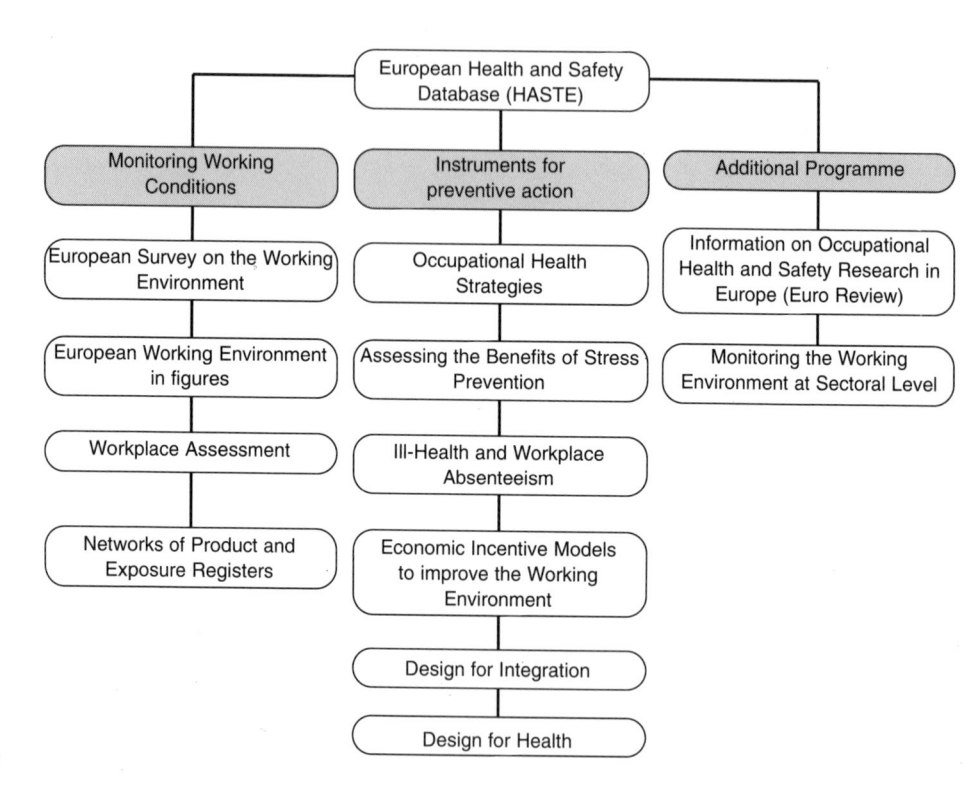

```
                    European Health and Safety
                        Database (HASTE)
                               │
    ┌──────────────────────────┼──────────────────────────┐
Monitoring Working        Instruments for           Additional Programme
   Conditions           preventive action
        │                      │                           │
European Survey on the     Occupational Health      Information on Occupational
Working Environment          Strategies             Health and Safety Research in
                                                       Europe (Euro Review)
        │                      │                           │
European Working         Assessing the Benefits       Monitoring the Working
Environment in figures   of Stress Prevention         Environment at Sectoral Level
        │                      │
Workplace Assessment     Ill-Health and Workplace
                              Absenteeism
        │                      │
Networks of Product and  Economic Incentive Models
Exposure Registers       to improve the Working
                              Environment
                               │
                      Design for Integration
                               │
                        Design for Health
```

Figure 5.5 The projects in the programme of the European Foundation for the Improvement of Living and Working Conditions relating to health and safety at work.

Database or statistics	Short description	Examples of possible use
GESTIS (dangerous substances)	The German "Berufsgenossenschaften" have a number of central databases on various topics. (Hauptverband der gewerblichen Berufsgenossenschaften,1990)	The data can be utilized by several persons and are useful in setting up workplace assessments. Gives data about dangerous substances
Accident report databases	Description of accidents, usually with (short) investigation reports	Prepare preventive measures
Accident and sick leave rates	Statistics of accidents and sick leave rates.	Benchmarking, comparison to other companies
BIA Handbuch	Catalogue of hazards and required precautions for dangerous equipment. The DIN 19 250 standard is used to indicate safety precautions. (Schütz & Coenen, 1985)	Find hazards and preventive measures for various types of equipment
HASTE	Catalogue of information systems for identifying risk factors affecting health and safety of workers and determining preventive measures. (European Foundation, 1993)	To find appropriate databases or statistics
SAFESPEC	Catalogue of hazards, possible consequences and safety objectives of workplaces and (dangerous) equipment. (TNO, 1990)	Find hazards and preventive measures for various types of equipment
Gefährdungskatalog Walzwerke	Extensive catalogue of hazards, safety objectives and standards, related to activities in iron and steel industry. (Hoor et al., 1993)	Find hazards and preventive measures for equipment used in steel mills

Table 5.1   Examples of databases and catalogues.

Sector studies
In many sectors, studies have been made about working conditions in that particular sector. As work processes and equipment are often the same or comparable within the sector, companies have mostly the same kind of problems. Because they take little effort, sector studies are a good start for workplace assessments, and are suited to small and medium-sized enterprises.

5.3 Evaluating results

Data collection leads to an overview of unwanted situations or situations that possibly lead to unwanted effects. As a next step, an evaluation must be made in order to find answers to the following questions:
- how serious are the possible hazards?
- should something be done about them?

Sometimes, data collection implicitly entails some kind of evaluation. For instance, in checklists some form of standard or reference is incorporated in the questions or checkpoints.

Using criteria
Safety and health criteria are effective means of evaluating hazards.
A comparison is made between the actual situation and the criteria.
Gaps observed should lead to further action until the actual situation meets the criteria. Meeting a standard implies in fact that any remaining risk is regarded as acceptable.
Criteria or company standards play an important role in the evaluation of hazards. Therefore criteria should be set with care.

Though it is possible to draw up ad hoc criteria for every hazard encountered, it is preferable to set criteria in advance (see section 4.7).
In good workplace assessment practice, criteria meet the following conditions:
- criteria or company standards are agreed on throughout the organization; All staff have been involved;
- criteria are written down, so reference can be made to legal

Topic	Subject	Sources of criteria
Working environment	- noise	national legislation, ISO 1991 NFX 35 108
	- climate	NEN-ISO 7726, NEN-ISO 7243
	- vibrations	ISO 2631, ISO 5349
	- lighting	NEN 3087, DIN 5035, ISO 8995,. NFX 35 103
	- radiation	various threshold limit values
	- gases, vapours and dust	various threshold limit values
	- biological agents	for toxins produced by micro-organisms: various threshold limit values
Workplace	- space, layout	ergonomic guidelines, national legal regulations, anthropometric data, NFX 53 102
	- furniture	DIN 4551, NEN 1812, NEN 2449, DIN 4549
	- controls, displays	NEN 3011, ISO 9355, DIN 43 062
	- VDU work	national legislation, ISO 9241, DIN 66 234
Physical workload	- manual material handling	NIOSH guideline for lifting, DIN 33 411, NFX 35 109
	- exertion of forces	DIN 33 411, NFX 35 106
	- heavy work	ISO 8996
	- postures	DIN 33 416, NFX 35 104, DIN 33 406
Safety		Various national regulations, various standards
Mental workload		ISO 10 075
Stress	- cycle times - control demands, control capabilities	incorporated in thematic assessment methods such as WEBA, VERA
Emotional strain		no criteria available
Work schedules, shiftwork	- working hours - shift work	national legislation

Table 5.2 Possible sources of criteria (the table does not give an exhaustive overview).

requirements, (inter)national standards, threshold limit values, ergonomic guidelines and the like;
- any company standard should have legal requirements as a minimum;
- top management are committed to the standards;
- criteria are not permanent, but are adjusted when necessary.

It is useful to set the most important criteria as a part of the company policy. For instance, a company may have as a policy the limiting of concentrations of gases and vapours to half the threshold limit value. Criteria can be described in a "Safety and Health Handbook". However, it is not possible to set up criteria for every possible hazard in the work situation. Others may not be realistic. In such cases criteria can be drawn up and discussed in the evaluation. Table 5.2 gives an overview of possible sources for criteria.

Risk assessment and calculations
Risk estimate techniques can give valuable information for discussing action plans. The first question about many hazards is: 'how serious is this, what consequences can be expected?' An assessment of risks, however simple it may be, supports decision-making on action plans. It allows an optimal allocation of resources.
As with all techniques and methods, the effort put into risk assessment must be in balance with the severity of possible effects.
A number of techniques or methods can be used for an evaluation of risks. In all cases, the starting point is a list of situations or activities that can possibly cause harm or damage. An overview of methods is given in table 5.3.

The result of a quantified assessment may be a number (for instance, 1 lethal accident to be expected in 10,000 man years of work). In itself this number is not enough to decide on an improvement. An important factor is the risk that is accepted or the risk people are willing to accept.
Four factors have an influence in discussions about the acceptability of risks:
- utility;
- aversion (such as risks of nuclear energy);
- time before effect becomes apparent (people are prepared to take

Instrument	Short description	To be used at
Formal methods (HAZOPS, FMEA, fault tree analysis)	Mostly complex and time consuming. Requires high levels of knowledge. Absolute figures are obtained, methods make no reference.	Suited for situations with high risks or severe consequences
epidemiological data	Powerful (if data are available). Some standards or threshold limit values are based on epidemiological data. Reference is made to other situations.	Evaluation of health hazards
standards, threshold limit values	If available, simple way of determining whether situation is acceptable. Gives little information about severity or probabilities. Sometimes based on epidemiological data. Reference is made to standards.	Evaluation of several aspects of work situations, especially exposure to chemical substances, vibrations and noise
DIN 19 250	Classification scheme for safety of machines. In Germany (BIA) a catalogue is based on this system. An absolute ranking is obtained, though the classification scheme is essentially the same as relative ranking.	Evaluation of safety hazards of equipment and machines
relative ranking techniques	Simple classification schemes for various kinds of risks. The number of classes varies. Mostly all factors in the risk calculation formula are scored. The outcome gives information of the risk and the priority an action for improvement should have. Ranking is relative.	Evaluation of safety and health hazards

Table 5.3 Overview of risk estimation techniques. Both quantitative and semi-quantitative techniques are summarized.

```
                                    W1  W2  W3
              S1
                                 1   -   -
                         G1
                             2   1   -
                     A1
              S2         G2
                             3   2   1
                         G1
                     A2      4   3   2

                         G2  5   4   3

              S3     A1      6   5   4

                     A2      7   6   5
              S4
                             8   7   6
```

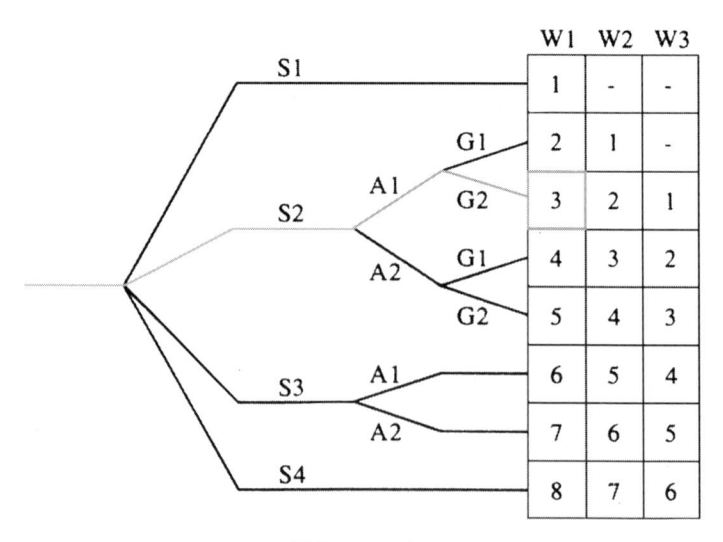

Risk parameters

Possible harm or damage	S1 = minor injury
	S2 = major, irreversible injury of one or several people or death of a person
	S3 = Death or several people
	S4 = many deaths
Exposure time	A1 = sometimes or more
	A2 = often to continuous
danger avoidance	G1 = possible under certain circumstances
	G2 = hardly possible
probability	W1 = very low
	W2 = low
	W3 = relatively high

	Safety requirements								
	1	2	3	4	5	6	7	8
Stochastic failures	*	*	*	+	+	+	+	+
Stochastic failures (very rare)							+	+
Systematic failures						*	*	+	+
Sudden failures	+	+	+	+	+	+	+	+
Failure due to drift						*	*	+	+
Failure detection	*	*	*	*	+	+	+	+

+ = take safety precaution   * = take safety precaution in some aspects

Figure 5.6 *Classification of risks and selection of actions according to the German DIN 19250 standard. For each piece of equipment the risk parameters are established. Following the path (grey line) leads to a category of safety requirements. In the table corresponding precautions are summarized.*

higher risks if effects are only long-term);
- voluntary behaviour (when risk taking is their own choice, people accept higher risks).

Comparison of standards can reveal whether there is a problem or not. A drawback, however, is that standards generally give no information on the severity of risks. Standards have been formulated for working conditions that can give serious problems, but also for more innocent hazards.
The way in which a semi-quantitative evaluation is made offers the opportunity to compare different hazards. For a number of safety and health hazards these so called relative-ranking methods are useful.

In the process of evaluation, risk communication and opinions play an important role. Full cooperation requires that workers have unbiased information about possible effects and risk levels and can form their own opinions.

Reports
Reports are essential in communicating the results throughout the company. A report fulfils several functions:
- information of management or stakeholders, not directly involved;
- basis for discussion and negotiation;
- milestone in the workplace assessment and improvement process;
- document for later reference;
- fulfilment of legal obligations.

In every workplace assessment process, reports should be drawn up. Timing, topics and the persons to whom the report is directed are dependent on the organizational context and the company's own practices and culture.

5.4 Selection of priorities

Conducting a workplace assessment may lead to an overview of hazards or risks. As mentioned, workplace assessment may never be a goal in itself or an excuse to postpone or put off actions for improvement. The results of assessments are the starting point for changes in the work situation.
In general a company's resources for carrying out and financing innovations

Example 5.3   Company safety and health reports (Germany).

Background
Sickness does not just mean absenteeism, personnel costs and organizational difficulties. It also signifies the sickness and health of the company. Higher sickness rates signal unfavourable competitive conditions. The sickness rate is often discussed in an antagonistic manner in companies. Numerous prejudices and incorrect assessments arise between health aspects and economic considerations. They often lead to the situation where the sickness rate is no longer discussed as a health problem of the employees but as an economic problem of the company.

Health reports
A company health report should provide the responsible persons in the company (primarily personnel management, the company doctor and the works council) with information on the distribution of work incapacity in the company. With the help of the report it is intended to facilitate the development of measures. The health report can become the central element of a discussion and decision-making process.

Contents of a health report
A company health report can have the following contents:

1. The company and its employees
   1.1 The company
   1.2 Company employees according to age, sex, occupational status and occupation
   1.3 The situation regarding health promotion in the company
2. The sickness rate in the company
   2.1 The sickness rate of the employees as compared with the situation outside the company
   2.2 The selection of examination or target groups for in-company health promotion
3. The health situation of the selected target groups for health promotion
   3.1 The target group according to age and occupational status
   3.2 The sickness rate in the target group
   3.3 An overview of the major health aspects of the selected occupations
   3.4 Selected major health aspects (subsection for each problem).
        - findings
        - possible causes
        - possible measures
4. Health promotion programme for the company
   4.1 result of the discussion in the company and in the working party
   4.2 Elements of a health promotion programme

(Source: Bundesanstalt für Arbeitsschutz, 1991).

are limited, so attention should be paid to the selection of priorities and action planning must be done with care. A number of arguments may play a role. First, priorities have to be right. One must clarify which situations have the greatest risk of severe harm for the largest number of persons. Three aspects can be taken into account:
1. seriousness of hazards or risks;
2. probability or likelihood;
3. number of persons exposed and duration of exposure.

Having made a priority based on the hazards or risks, one has to decide on the possibilities of dealing with the risks and taking steps. Aspects that play a role are:
4. causes of problems.
5. possible options;
6. feasibility of solutions;
7. opportunities in investment planning.

In this section, each aspect will be discussed briefly.

1. Seriousness of hazards or risks
Estimating risks and gravity of consequences can be a good method of creating a ranking in the hazards of risks encountered (see section 5.3). It is usual to take action on the highest risks first and to pay minimal attention to trivial risks. An advantage of this method is that an objective statement of priorities can be generated. This method is especially useful when it is possible to estimate risks (in either an absolute or a relative sense). At first glance, relative ranking or risk assessment may offer good opportunities for setting up criteria. However, some drawbacks and limitations must be noted:
- Risk estimation techniques mostly require profound knowledge of occupational safety and health. In general, quantified risk estimation is an expert's job.
- Some topics are hard to quantify. For instance stress risks are difficult to ascertain precisely, as they can vary from one person to the other.
- Risk estimation techniques only deal with observable hazards and risks. The opinions and complaints of workers are generally not dealt with.

Note that in some definitions of "risk" the severity of the consequences is incorporated.

1. harm consequence grid

	high	medium	low
harm to people (accident)	major injury/accident resulting in over 3 day absence	hospital visit and/or absence for up to 3 days	up to and including first aid
harm to plant (accidental damage)	accident resulting in long term damage costing time and money	accident resulting in short term damage	slight damage
harm to environment	large scale emission to air of toxic gas	small scale emission to air of toxic gas, not significant	unplanned emission of non toxic substance

2. exposure factor (multiply factors (a) and (b))

(a) Frequency of exposure					
0	1	2	3	4	5
no exposure	fairly low	low	medium	fairly high	high exposure

(b) number of persons exposed					
0	1	2	3	4	5
none	1 - 5	6 - 10	11 - 20	21 - 49	50 +

3. categories of risk/possible actions

Category	Grid reference	Exposure factor (0-25)	Action priority, moderated by exposure factor
A	high likelihood high consequence		Immediate action, high likelihood of serious injury
B	high likelihood medium consequence		Immediate action, reduce likelihood
C	high likelihood low consequence		Seek longer term means of reducing likelihood, may need to judge priority
D	medium likelihood high consequence		Plan reduction of likelihood of event. Consider design of lower consequence system. Judge priority
E	medium likelihood medium consequence		Plan reduction of likelihood of event. Consider design of lower consequence system. Judge priority
F	medium likelihood low consequence		Judge priority, long term plan to reduce likelihood
G	low likelihood high consequence		Monitor standards regularly to reduce, maintain likelihood to lowest possible level. Seek design of lower consequence system
H	low likelihood medium consequence		Monitor to maintain standards. Consider the possibility of lower consequence systems
I	low likelihood low consequence		Monitor annually to ensure likelihood does not increase

Figure 5.7   Relative ranking techniques for prioritizing risks. Sequence for determining whether action is required. The three tables can be evaluated in sequence. (Source Loughborough University, 1992).

2. Probability or likelihood

The likelihood of an adverse effect occurring is a factor that should be taken into account. The greater the probability, the greater the need to take measures.

3. Number of persons exposed and duration of exposure

The more persons exposed to hazardous situations or activities, and the longer the duration of exposure, the more elimination or reduction of hazards is needed. One should note that often more persons are exposed

Cause analysis/survey work   **what takes place over time**

	need for control	real cause	cause in the work process	factor causing damage	sequence of events	injury, illness
A	Programme for working environment	Job factors	Condition of the workplace	Accidents: the damaging factor at the moment of the accident	Accidents: the atypical event at the moment of the accident	The person
B	Standard for working environment	Personal and social factors	Execution of the work	Work-related diseases: the damaging long-term effect	Work-related diseases: the normal long-term sequence	The production equipment
C	Standard observance					Raw material/ product
D						The environment

Table 5.4   Injury/illness-cause model. It is intended to approach the working environment from the control position (left side of the table) and not from the damage position (right side). (Source: Øland, 1993).

than is obvious. For example, noisy tools lead to exposure for every person present in the workplace, not just the user of the tool.
In general: the higher the number of persons exposed, the higher the priority for reducing the risk.

4. Causes of problems
To reduce risks it is essential to find the principal underlying causes. One should avoid a mechanistic way of problem-solving by focusing too much on the symptoms. Often multiple causes contribute to a single hazard. Problems are best solved by eliminating the most fundamental cause, therefore profound knowledge of the work process and the way it is actually performed in practice is essential.

It is often felt that accidents could have been avoided. Better preparation, better organization, better coordination between the objective of total quality with control of economic and social cost, can contribute towards prevention. However, numerous communication defects on the whole line, from design to realisation of a work situation, also contribute to accidents. A useful way in which to make a systematic overview of possible causes of a different nature is the so-called fish bone diagram.

In safety matters, there is no single cause-effect relation. A number of factors, effects which stem from design to operation, can culminate in an accident. Thus managers should look for a global approach, which will involve all the actors.

5. Possible options
In most situations, risks can be reduced in a number of ways. For instance the risk of hearing loss can be reduced by personal protective equipment or by using less noisy machines. From an occupational safety and health point of view the hierarchy listed below should be used:
1. Elimination of the hazard, for instance by substituting dangerous chemicals by less hazardous substances
2. Isolation of the source: put noisy machines in a sound-proof cabinet, use of closed vessels
3. Measures in the environment: ventilation, noise absorbing materials

4. Isolate workers from the process, use of isolated (control) rooms
5. Organizational measures: reducing exposure time, dividing strenuous activities among more people, better planning
6. Use of personal protective equipment.

6. Feasibility of solutions

Because workplace assessments are only one step in the process of improvement of work situations, one must look ahead and evaluate options for improvement beforehand. It is no use considering extensive solutions if financial resources are limited. The feasibility of possible solutions is determined by a number of aspects:

= financial or economic aspects

A cost-effectiveness analysis of proposed improvements can be performed. Though many benefits of working conditions improvements cannot be stated in terms of money, balancing costs and intangible or non-financial assets may provide important information on feasibility. At least one can say if the benefits are worth the cost of an improvement (see example).

= technical feasibility

Some problems cannot be solved with current technology

= operational feasibility

Some solutions proposed may interfere with operational procedures within the company. If, for instance, a lifting aid slows down frequent handling actions too much, it will be set aside. In many cases solutions proposed by external occupational safety and health experts have poor operational feasibility. The only way to deal with this is to have workers involved.

Each of the intended measures must fit into company policies.

7. Opportunities in investment planning

The best time to carry out improvements is when major investments or reorganizations are planned. Investments offer the opportunity to solve certain occupational safety and health problems in a relatively cheap and effective way. In certain industries, the changing of a product also means

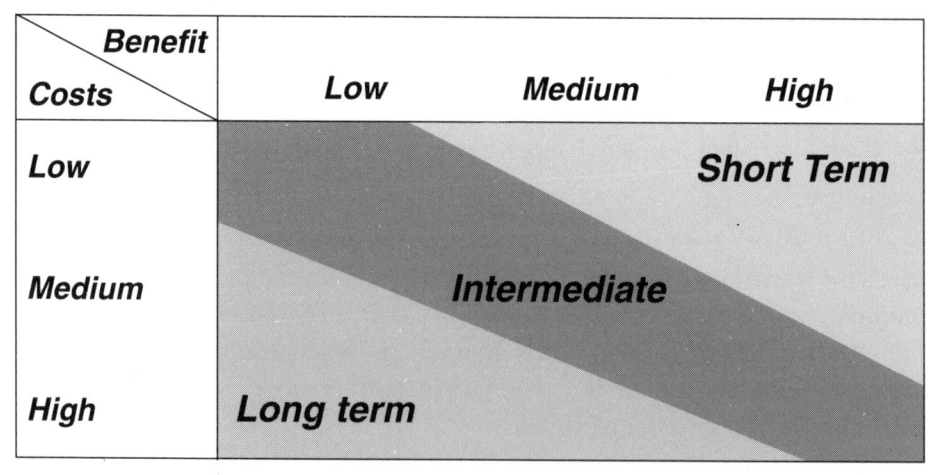

Costs \ Benefit	Low	Medium	High
Low			Short Term
Medium		Intermediate	
High	Long term		

Figure 5.8 Costs-benefit analysis of possible measures as a tool in action planning.

changes in the production line. These opportunities should be used. It may be useful to adjust priorities to the opportunities offered by investment planning.

5.5 Action planning

Action plans are not solely the result of an evaluation of risks. Other considerations, such as keeping the improvement process going, are also important. Sometimes it is better to carry out a small and easy to implement improvement first, and make the larger changes at a more suitable time.

nr.	hazard	measures	person	deadline	budget

Figure 5.9 Sample action plan.

In its simplest form a good action plan involves
 (i) an inventory of problems and situations, to be dealt with,
 (ii) the name of a person who takes care of carrying out required measures,
 (iii) a time line
 (iv) the budget.
In order to be effective, an action plan has to be evaluated on a regular basis (see also chapter 6).

Design and intervention projects
Design and intervention projects are closely linked to workplace assessments in two ways. Firstly, a workplace assessment may lead to improvements for which a design must be started. Secondly, design projects can be initiated by some other need (for instance modernization).

Improving Performance 6

6.1 Introduction

Once workplace assessment has been carried out, a need for further and systematic improvement may arise. A company can develop from an ad hoc attention to safety and health (for instance after an accident) to a situation in which attention and improvement is structural and part of the company culture. One way to achieve this is by setting up a safety and health management system. The purpose of such a system is:
- to control and to reduce risks to acceptable levels;
- to reduce uncertainty in risk decision-making;
- to ensure that safety and health are adequately dealt with;
- to increase confidence in decisions on the improvement of working conditions.

The last is meant primarily for workers in the company, but it has also an external function. Workers are given confidence that safety and health is part of daily management practice and that working conditions are considered whenever necessary. A safety and management system also

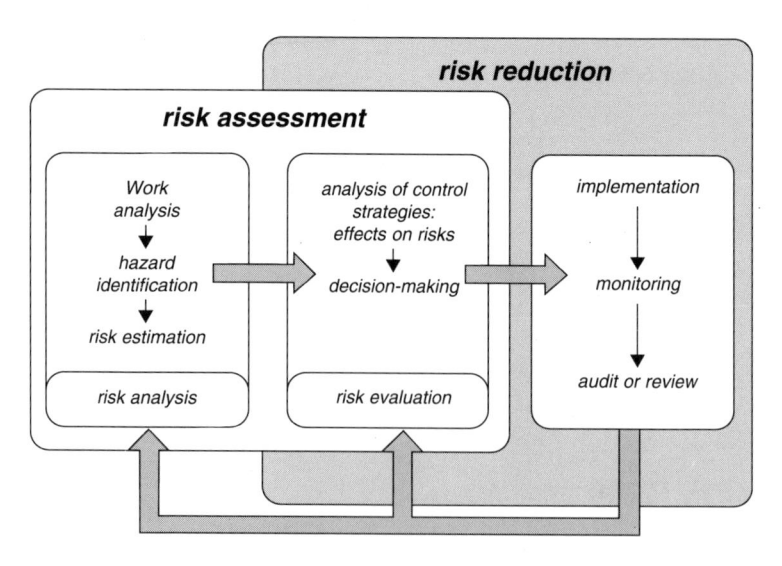

```
                    risk reduction
      risk assessment

  Work          analysis of control    implementation
  analysis      strategies:
    ↓           effects on risks            ↓
  hazard              ↓
  identification   decision-making       monitoring
    ↓                                       ↓
  risk estimation
                                         audit or review
  risk analysis    risk evaluation
```

Figure 6.1   Overview of a risk management system. (Source: Cox & Tait, 1991).

assures authorities (such as the Labour Inspectorate) that legal obligations are being met and that all reasonable actions to prevent accidents and harm to health and well-being have been taken. In some countries (Sweden, Holland, U.K. Norway) it is a legal obligation to set up some form of 'internal control'.
Furthermore a good functioning safety and health management system gives the company a good image. This may improve the competitive position of the company on the labour market.

A complete safety and health management system can have extensive performance objectives, spread over a large number of the firm's activities. In example 6.1 an elaborate set of performance criteria is given. This system is similar to the intentions and the system of quality control according to the ISO 9000 standards.

6.2   Measurement and review

Executing workplace assessments and setting up improvement projects require an investment of time, energy and financial resources. For management, it is useful to know if these assets have been effectively used.

Example 6.1   Objectives of performance standards.

Control of inputs
Design and selection of premises
- safety and health (S&H) aspects are included in plans to build or buy premises;
- S&H aspects of construction are considered
- S&H are specified in contract specifications.

Design and selection of plant and substances
- all relevant issues are considered at the design stage;
- all design specifications refer to S&H requirements and are specified in documents;
- S&H performance is considered in selecting contractors;
- all relevant data are collected;
- procedures for receipt and storage of goods ensure safety and health criteria are checked;

Plants and substances used by others
- standards should ensure that plants and substances used by others are appropriate;

Acquisitions
- standards should ensure that health and safety are considered in business acquisitions;

Human resources
- standards should ensure that employees are recruited on the basis of relevant selection criteria;
- criteria are also based on assessments and health and safety analysis;
- health and safety performance is considered when awarding contracts.

Information inputs
- all relevant information is gathered.

Control of work activities
Control
- necessary organization and procedures are maintained for policy, organizational design, planning and reviewing;

Cooperation
- involvement of all working people is ensured;

Communication
- flow of all relevant information is ensured;

Competence
- all workers are competent with respect to safety and health aspects of their work;

Performance standards for risk control
- standards should ensure that risks are eliminated. Included are: normal operation, maintenance, planned change, foreseeable emergencies.

Control of outputs
Products and services
- standards ensure that products are designed to ensure health and safety in use, storage and transport;
- arrangement should be made for packaging and labelling to ensure safety and health;

By-products of work activities
- risks to others who may be affected are considered;
- appropriate control of unwanted outputs is ensured

Information for external use
-standards ensure the compilation of information for safety in connection with purchase, use, maintenance, handling and storage;
-information is compiled for non-employees that may be affected.

(Source: HSE, 1993)

Results must be apparent. Measurements, or a renewed assessment, can show some results (for instance lower noise levels or fewer complaints). Some results, however, may not become apparent at once:
- a low accident rate (even over a longer period) is no guarantee that risks have been reduced and are effectively controlled;
- sick leave rates may not change immediately after preventive action has been taken.

In order to ensure the effectiveness of a safety and health policy, continuous monitoring is necessary. Various methods can be used:
- performance monitoring of managers, check on activities;
- periodic examination of documents (e.g. action plans);
- regular inspection of premises (especially with respect to safety measures);
- reporting incidents and accidents;
- continual monitoring of sick leave rates;
- labour turnover;
- productivity.

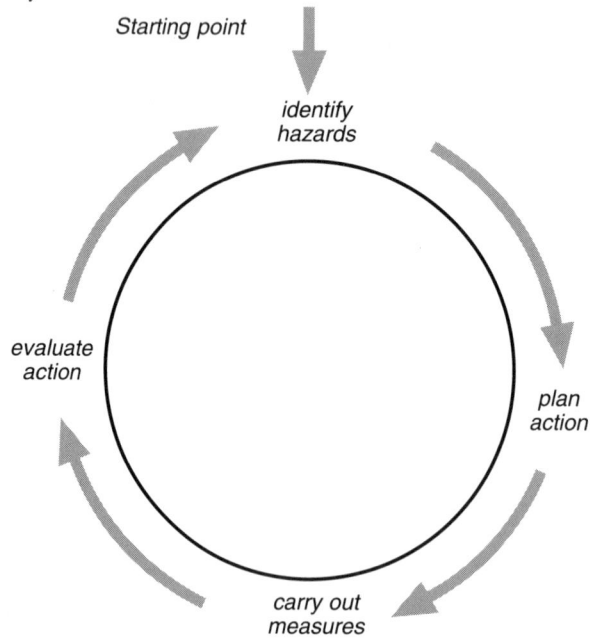

Figure 6.2   Four step safety management system. (Source: Ministry of Social Affairs and Employment, 1994).

The ability to learn from experience is of the utmost importance in every aspect of organizational performance. The same goes for safety and health management. A review of the workplace assessment process helps the company to find weak points and to improve its way of working. A review can focus on a number of topics, for instance:
- goals; have goals been realistic?
- staffing; were all persons needed involved?
- the quality of the assessment process; has the assessment been carried out according to accepted standards?
- planning; was planning realistic, why were deadlines not met?
- worker involvement; have policies been negotiated, has sufficient information been provided?

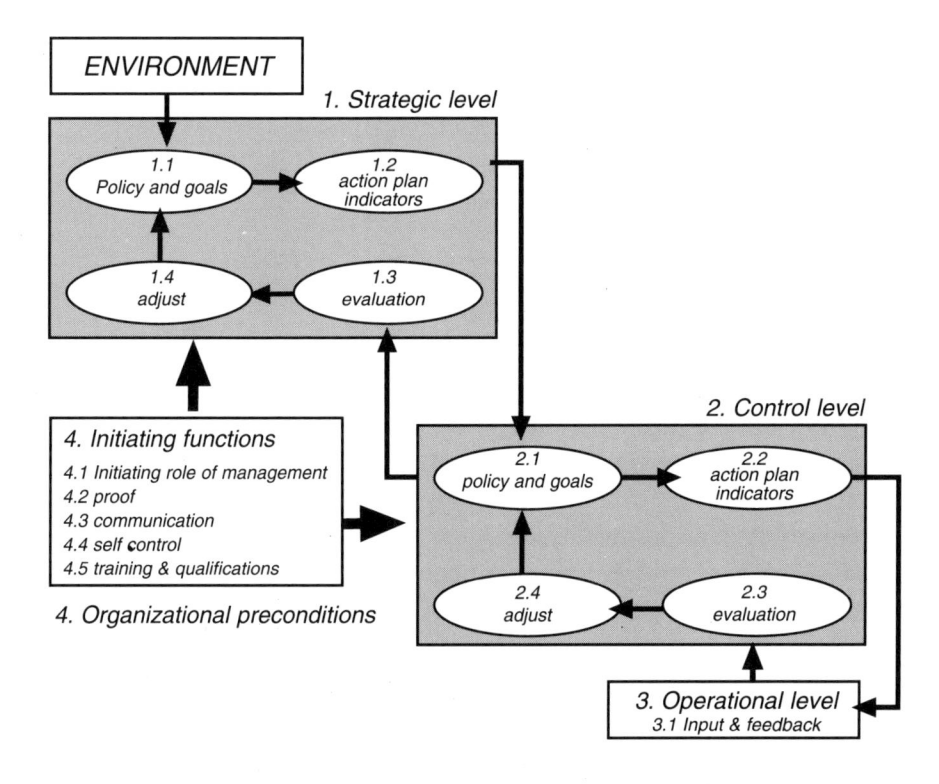

Figure 6.3   Three level safety and health management system (Source: Van Hezik, 1994).

- adequacy of measures; have measures been focused on the most important hazards?
- performance of responsible persons; have persons taken an active attitude and have tasks adequately been carried out?
- cost - effectiveness ratio; have improvements been effective, is the cost in balance with the reduction in risks?

In addition to this list, a number of other topics can be reviewed.

6.3 The management cycle

The management cycle is a general concept which offers good opportunities for ensuring that workplace assessments lead to improvements. In fact, a management cycle incorporates all activities that are needed to get a continual improvement. This way, an internal control system is created. One of the outcomes is that the company's self activity is enhanced.

Basically, it is a feedback control system in which the actual situation is compared to a desired situation. The gap between the two is closed by improvements in working conditions. An occupational safety and health management cycle may consist of four steps, based on the Deming circle (plan, do, check, act). In practice, systems with any number of steps are possible (see figure 6.2):
- identification and evaluation of hazards (measurement of the actual situation and comparison to the goals);
- design of improvements (draw up plans for corrective action);
- implementation of improvements (carry out plans);
- review of the effect, evaluation (in fact, the review is a renewed measurement of the situation).

Many control systems for quality and environmental issues are built on a three level system, each level representing part of the organizational functions:
- strategic
- control
- operational.

Mostly, every level is elaborated in handbooks in the form of procedures (control level) and work instructions (operational level). The three-level system is illustrated in figure 6.3.

Example 6.2 *Tasks and responsibilities with respect to internal control of safety and health in Norway.*

Managers
- define objectives;
- define safety requirements;
- ensure continual assessment of the enterprise's routines and products;
- work out strategies and progress schedules for attaining objectives;
- ensure that all employees receive good information about objectives, strategies and plans;
- require subordinate managers to accept responsibility for health, environment and safety.

Middle management, work supervisors
- accept a clear management responsibility for health safety and environment;
- make an active effort to learn about the experiences of, and seek assistance from, safety delegates and health personnel;
- systematically obtain key figures about safety, health and environment at regular intervals;
- take initiative and become a driving force for internal control efforts.

Safety delegates
- receive training (and the required time) in internal control and solving problems;
- participate in regular inspections, together with manager.

Safety and health personnel
- systematize their findings from visits to workplaces and health examinations, provide intelligible information;
- ensure that information is considered in relation to data on sick leave, turnover and branch statistics and point out necessary measures;
- work to ensure that health, environment become an integrated aspect of the activities of the enterprise.

Employees
- show responsibility for their own safety and that of others, make an active effort to reduce risk;
- make an effort to ensure that their working experience is properly used (also in connection with health, environment and safety);
- follow the established instructions.

(Adapted from: Ministry of Local Government, Oslo, 1991)

Example 6.3   Total Safety Management at ABB

At Asea Brown Boveri a Total Safety Management System has been introduced. The system is amongst others based on EU directives.

Goal
Lowering work-related failures and lowering accident rate.

Basis
Work safety act (German), EU directives.

Strategy
TOP-programme for improvement of occupational health: integral consideration and design of:
- technique (T);
- organization (O);
- personal working conditions (P);

Method
1. Selection of working situations for assessment of failures, occupational hazards, sick leave rates etc.
2. Setting up of occupational health circles in every work area, consisting of 4 workers, foreman, works council representative, plant manager, medical officer, safety officer (acting as moderator).

Preparation
1. Creation of commitment and trust.
2. Top-down information/training on the framework of safety and psycho-social working conditions.
3. Translation of methods for safety management to work in small work groups.
4. Translation of methods for hazard identification to local and individual use.
5. Definition of working conditions (physical, chemical).
6. Definition of subjective factors.

Expectations
Realisation of goals, plus:
- ergonomically improved working conditions;
- improvement of cooperation;
- development of ability to give positive and constructive criticism;
- improvement of appreciation of personal safety and health;
- more useful proposals for improvement;
- testing of work methods;
- better identification with own workplace;

Initiation
Task force consisting of human resources manager, works council representative, medical officer and safety expert.

(Source: Huf, 1993)

On a company scale, this cycle can be repeated on a yearly base, or even less frequently. At a workplace or department level, more frequent cycles are possible. Each time something is changed (for instance new machinery) a complete cycle can be carried out.
Eventually, the use of management cycles can evolve into a practice of continual improvement, like the "Kaizen" system. In this practice workers and management permanently optimize products and production methods. Organizational learning is a prerequisite.

Tasks and responsibilities
A management system may evolve into a centralistic and bureaucratic way of working. The best way to overcome this danger is to integrate the cyclic way of working into everyday practice. Workers must have the responsibility and opportunity to notice (and communicate) problems, and cooperate in the development of solutions. The evaluation is very much in the hands of the workers. Setting up safety and health circles also offers good opportunities for a bottom-up approach in safety and health management.
It is of importance that line management and staff should commit themselves to a non-bureaucratic way of working, and promote and support it. Management must coordinate actions throughout the company.

Safety management systems are one way of establishing internal control of working conditions. In Norway and Sweden companies are obliged to have an internal control system. In example 6.2, tasks and responsibilities concerning safety and health are listed.

Small and medium-sized companies
The concept of management cycles, and especially the documentation of each of the steps, is based on practice in large companies. With respect to internal control in small and medium-sized companies, some remarks can be made (Johannson & Johannson, 1993):
- Work is carried out in a result-oriented way. Freedom of choice in the manner of achieving goals is important for company owners.
- Solving problems and planning is informal.
- In small companies the cost of a formal way of working is relatively high, whereas benefits are limited.

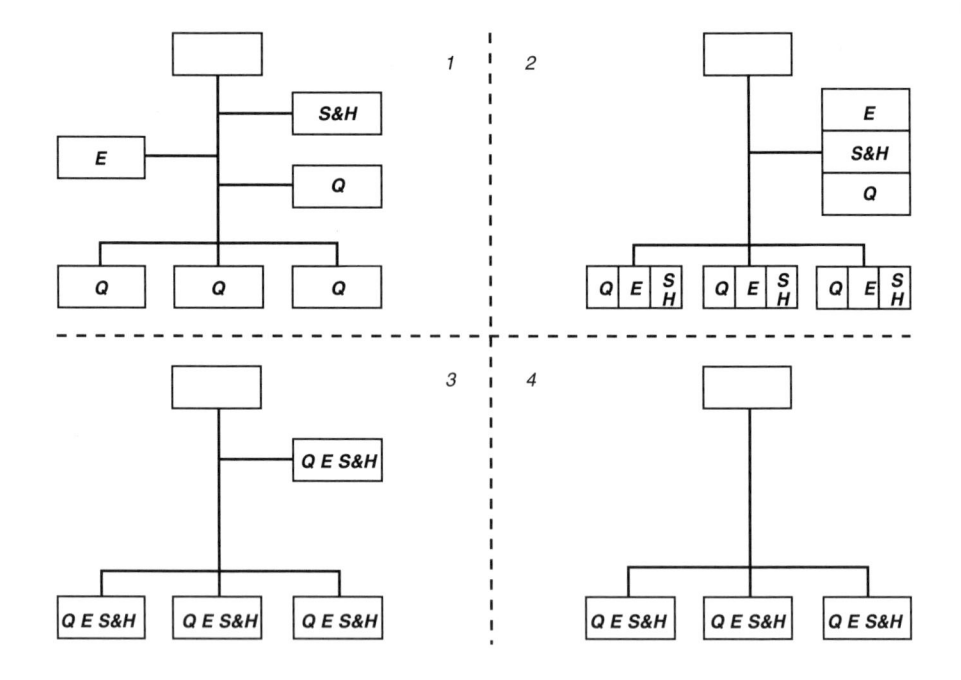

Figure 6.4 Possible options for organizing integrated quality, environmental and safety and health control (1 = separated aspects, limited quality control in line management; 2 = aspect control, quality, environment and safety & health are separated; 3 = integrated support by staff, integral control in line management; 4= integrated control in line management, support by external services).

6.4 Introducing safety and health management

Commitment
Safety and health management systems can have a great influence on the everyday functioning of companies. Every member of the organization should at least check his/her actions against possible consequences for working conditions. For instance, designers or engineers have to deal with ergonomics, purchasers have to check materials or tools for safety aspects and so on. In most organizations, such changes are difficult to establish. Of paramount importance, however, is a commitment to structural attention to working conditions. Having a formal system of procedures and work instructions is helpful, but is of secondary importance.

> **Example 6.4** *Organization and structure of combined quality, occupational safety and health, environment and organization management system.*
>
> At Deutsche Tiefbohr-AG (DEUTAG) a combined management system has been created from three existing systems. Each system has its own handbooks in which procedures are described. All three systems have similar structures.
>
> **Quality management system**
> The handbook comprises the complete quality control system, based on ISO-9000.
>
> **Health, safety and environment management system**
> The HSE handbook describes a system of instructions and measures.
>
> **Organization management system**
> Description of administrative procedures.
>
> *policies & objectives*
> *procedures*
> *instruction manuals*
>
> quality | health, safety & environment | organization
>
> source: Altmann, 1993

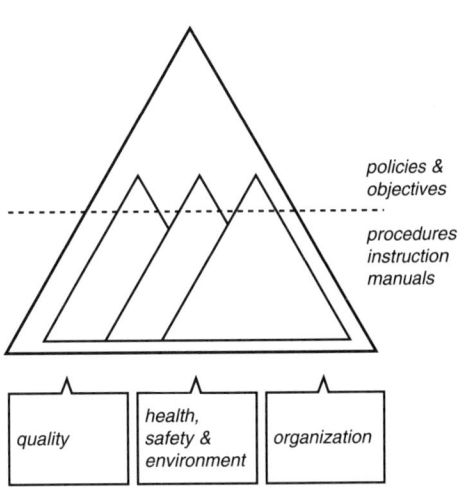

Links with other aspects of company policies
Linking management systems for working conditions, environment and quality control may offer some advantages:
- As systems are based on the same principles, experiences with quality control or environmental issues can be used to set up a similar way of working for occupational safety and health.
- Working conditions and environments have common issues, such as toxic or dangerous chemicals and noise. These can be handled at the same time, by the same person and with the same methods.
- Comparable systems for quality, environment and working conditions can improve cooperation between these fields of activity.

In practice, some problems have to be tackled:
- The relative importance of quality, environment and working conditions

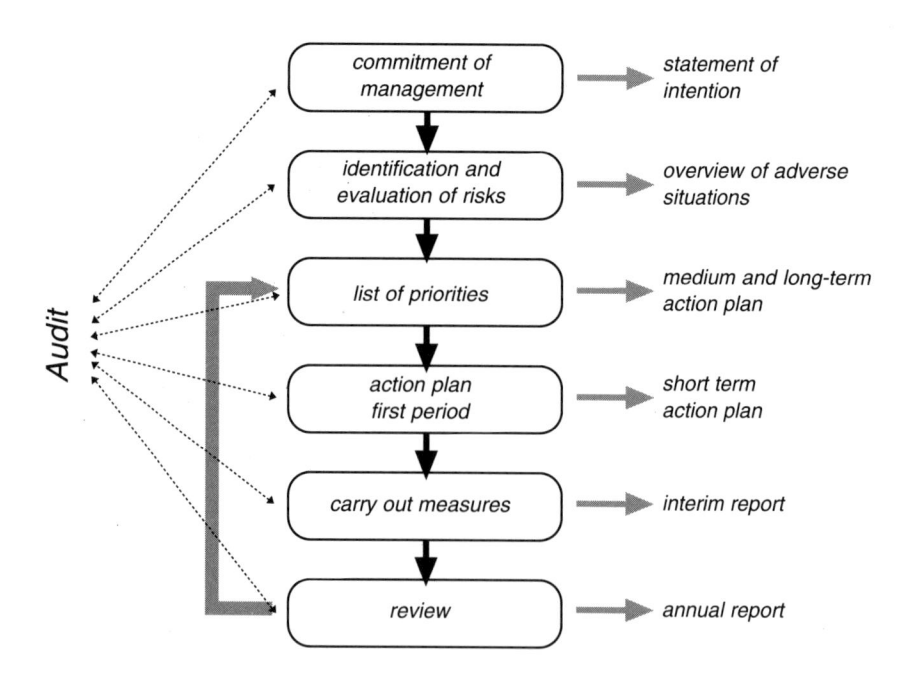

```
Audit ←------→  commitment of        ──▶  statement of
                management                intention
                     │
                     ▼
                identification and    ──▶  overview of adverse
                evaluation of risks        situations
                     │
                     ▼
                list of priorities    ──▶  medium and long-term
                                           action plan
                     │
                     ▼
                action plan           ──▶  short term
                first period               action plan
                     │
                     ▼
                carry out measures    ──▶  interim report
                     │
                     ▼
                review                ──▶  annual report
```

Figure 6.5 Audit as a part of safety and health management systems (Adapted from Lindsay, 1991).

Instrument	short description
CHASE	Checklist to be filled out by management. Based on scores on questions, the profile will show up strong and weak points in the safety and health management system. Two versions are available (for companies with less or more than 100 employees).
OSART	Audit instrument for nuclear industries. The company has to provide a complete description of its organization. Research and interviews may take 2 to 3 weeks.
MANAGER	Audit system for complex chemical industries. Meant as an addition to quantified risk assessments.
BCISC	General (open) checklist. For each company additions or specification of the checkpoints is required.
ISRS	Questionnaire with four categories of questions, including a professional judgement (to be made by the auditor). Requires trained auditors.

Table 6.1 Overview of audit instruments for safety.

Example 6.5   Overview of the contents of the OSART questionnaire.

The OSART audit instrument is meant for nuclear industries. Most of the topics also apply to other activities. Topics that are specific to the nuclear industry are omitted from this example.

1. Management
1.1   Structure of the operational organization
1.2   quality programmes
1.3   legal obligations
1.4   industrial safety
1.5   fire safety
1.6   operational safety
1.7   control of documents

2. Training and qualifications
2.1   organization of training
2.2   materials and facilities for training
2.3   training programmes
2.4   training for various groups of personnel

3. Operation
3.1   organization of administrative operations
3.2   procedures and documentation
3.3   process information and administration
3.4   facilities and equipment

4. Maintenance
4.1   organization of maintenance
4.2   maintenance personnel and their qualifications
4.3   maintenance program
4.4   maintenance procedures
4.5   preventive maintenance
4.6   corrective maintenance
4.7   inspection
4.8   management of contractors
4.9   facilities and equipment
4.10 storage

5. Technical support
5.1   Supervision of test program
5.2   process control
[....]
5.5   feedback of experiences
5.6   modifications

6. Radiation
[....]

7. Chemistry
[....]

8. Emergency procedures
8.1   organization of emergency procedures
[....]
8.5   procedures for introduction of emergency procedures
8.6   communication
8.7   facilities and equipment
[....]
8.10 training
8.11 public information
[....]

(Source: Oortman Gerlings, 1990)

may vary. Quality control generally has higher priority, and extra time for working conditions may be hard to find.

Many ways of organizing are possible. The use of specialized staff services (figure 6.4, option 1) has the advantage that scarce knowledge and experience is used to the best advantage. A drawback is that central decision-making may be slow, and active involvement of workers is harder to ensure.

If a company has experience with quality control systems, the same control systems can be used to deal with working conditions. It is possible to structure safety and health management according to quality control systems (see example 6.1 and 6.2).

6.5 Investigating performance: the audit

In general the aim of a management audit is to find out if an organization is capable of executing and actually performing all activities related to management cycles. In short, the purpose of the audit is to find if the safety and health management system works in an effective and efficient way. There are two forms:
- Internal audit: the occupational safety and health system is audited by the company itself in order to find weak points and to improve these. As a result the performance and the efficiency can be improved.
- External audit: the complete management system is evaluated on its efficacy by an external organization. This may be an inspectorate, an insurance company (e.g. for safety matters) or organizations that issue certificates (like ISO 9001 for quality control).

Essential to the audit is that data are collected in an independent way. In internal audits especially, one must ensure that the roles of client and auditor are clearly separated.

Just as in workplace assessments, the purposes and context of an audit must be clear. The scope and organization of an audit should be considered and planned in the same way as assessments.

Tools
A number of tools are available, mostly aimed at major safety hazards. So far, very few complete occupational safety and health audits have been developed.
Most of the instruments must be used in combination with training of auditors. In table 6.1 some instruments are briefly described.

Summary and Guidelines 7

Workplace assessment is a systematic investigation of all aspects of work. The aim is first to identify situations or activities that may cause accidents or lead to illness or discomfort. An evaluation is then made of whether preventive or corrective action is needed.
Subjects for workplace assessment include the working environment, chemical and biological agents, physical and mental workload, ergonomics, safety, stress, working hours and emotional strain.

Workplace assessment is the start of a preventive policy aimed at better working conditions that will benefit the company as a whole (for instance through fewer accidents and occupational diseases, better productivity and quality). Additional benefits for workers can also be attained (for instance better health, job satisfaction).

For companies that are beginning workplace assessment, careful organization is important. The process of workplace assessment should fit into company culture and should be performed in full cooperation with the workers. It is extremely useful to define a company policy that links workplace assessment to other organizational issues. In this way the assessment also serves other purposes.

Topics to be dealt with during the planning are, amongst others, the goals of the assessment, staffing and the role of experts, worker participation, the time-table and the follow-up.

For actually carrying out an assessment various methods and techniques can be used. Four types of assessments are discussed:
- Global inspection of work situations by means of checklists and observation. These are especially useful for companies that want to get a first impression and have little experience.
- Task-centred assessment focuses on the tasks as actually performed by workers. These are useful in design projects and in assessing activities.
- Thematic assessments focus on a single aspect of working conditions. The goal is to obtain detailed information. Most methods must be performed by trained assessors.
- Organizational or strategic assessments investigate the organization of occupational safety and health management.

The actual assessment has roughly two parts. The first is to identify situations or activities that can possibly be harmful. This part merely consists of collecting information from all kinds of sources inside or outside the company. The second part is an evaluation which must lead to a decision about possible measures to be taken. The result can be some form of action plan.

In the process of workplace assessment, follow-up (in the form of implementation of improvements and a review of the achieved results) deserves special attention. Assessments are not a goal in themselves and most of the benefits of workplace assessments are as a result of the improvement measures.

In an ideal situation working conditions are an integrated part of day-to-day management practice. It is the concern of every department and member of the organization (internal control). Structures of quality control (for instance the ISO 9000 standard) can be adapted for occupational safety and health management as well. By reviewing and auditing, occupational safety and health management is continuously improved.

Some guidelines can be of help in setting up successful workplace assessments.

Guideline 1. Commitment is essential
Commitment at top level is essential, to see that intentions for workplace assessment are clearly stated. Middle and lower management have to take their responsibilities and stimulate attention for working conditions and workplace assessment.
If commitment is lacking, point out how a company can benefit from workplace assessment and occupational safety and health management.

Guideline 2. Define a policy that links to other organizational issues.
See that a company policy on working conditions (including workplace assessment) is coherent with other organizational issues such as quality, environment, technological change and human resources management. Working conditions should be everyone's concern. Adopt the internal control principle as much as possible.

Guideline 3. Seize opportunities to start workplace assessment
Look for opportunities to start a workplace assessment. Good opportunities are:
- employees with complaints about working conditions, or high sick leave rates because of adverse working conditions;
- the company wants to start preventive policies;
- a design or investment project is started, or the company plans for a reorganization;
- an incident or accident has occurred;
- there is a change in legal obligations.

Guideline 4. Ensure adequate knowledge
Arrange for sufficient knowledge in order to
- be able to discuss with experts and consultants;
- perform as much of the workplace assessment as possible by employees of the firm;
- enhance the quality of safety and health management as a whole and workplace assessments in particular.

Guideline 5. Plan workplace assessments with care
Plan carefully. Note that time spent in organizing will be compensated for during the assessments and follow-up. Pay attention to:
- formulation of clear and realistic goals and objectives;
- adjustment to the company policy;
- methods to be used;
- staffing (training, tasks and responsibilities), role of external experts;
- worker participation, role of works councils, worker representatives and trade unions;
- time schedule;
- follow-up: selecting priorities, drawing up an action plan and design and implementation of improvements.

Guideline 6. Select suitable methods
Choose methods or instruments that provide sufficient information for reasonable effort. A multiple method approach often gives the best results.

Guideline 7. Ensure worker participation
Involve workers or their representatives. Workers should have an influence in working conditions policy and objectives (strategic level). They should also play a role in planning and performing assessments (operational level). Workers must be informed about and able to comment on the contents of reports. Planning follow-up (design and implementation of improvements) requires worker involvement as well.

Guideline 8. Learn from experience
Review the complete process of workplace assessment. The goal is to do better next time. A review comprises several activities, amongst others:
- to find weak points and less than optimal results;
- to make a cost-effect analysis;
- to establish whether all persons involved have carried out their tasks in an adequate way.

Guideline 9. Pay attention to working conditions in daily practice
Make workplace assessment and safety and health management part of everyday management practice. See that working conditions are given attention and are balanced against other subjects in terms of company management.

Glossary 8

With respect to workplace assessment, some concepts have different meanings in different countries. As an illustration some definitions are given.

Accident
All undesired circumstances that will give rise to:
- ill-health or injury;
- damage to property, plant, products or the environment;
- production losses or increased liabilities.

Assessment
Entire systematic process in which possible problems are identified, investigated and evaluated.

Assessor
Person or coordinated group of persons carrying out the assessment.

Audit
Structured process of collecting independent information about the way

safety and health management is structured and put into practice. Topics are efficiency, effectiveness, and reliability of the management system.

Cascade approach
Approach to the identification of possible hazards in which the first step is a global check. More detailed investigations are made if necessary.

Hazard
The intrinsic property or ability of something (materials, equipment or activity) with the potential to cause harm. A hazardous situation is a situation in which a person interacts with the hazard, but is not necessarily exposed to it. A hazardous event is the trigger which exposes a person to the hazard.

Incident
All undesired circumstances and 'near misses' that have the potential to cause accidents, or that could have led to an accident.

ISO 9000 series
Standards for the management of quality control. The standards consider every aspect of the organization that has an influence on quality.

Monitoring
Systematically measuring aspects of working conditions over a long period of time, in order to find changes.

Review
Description of activities involving judgements about performance and decisions about improving performance.

Risk
The likelihood that the potential for harm of a particular hazard will be attained under the conditions of use and/or exposure and the severity of the consequences.

Risk Assessment
The process of evaluating the possible extent of harm from a risk to the health and safety of workers while at work arising from the occurrence of a hazard at the workplace.

Risk Management
Coherent set of policies and activities used to increase safety and minimize losses, involving the identification, evaluation and control of risks.

Strategic or organizational assessment
Assessment of the organization of working conditions by management.

Task-centred assessment
Assessment that takes the tasks and activities performed by workers as a starting point.

Thematic assessment
Assessment focused on one aspect of working conditions, for instance stress, physical workload, or exposure to chemical substances.

Working conditions
The circumstances under which work has to be performed, or which influence the work. Working conditions cover the work environment, physical workload, mental workload, the psycho-social context and the organizational context.

Workplace
Location where a person normally performs his/her tasks or spends a reasonable time working.

Workplace Assessment
Workplace assessment is a systematic investigation of work, in all its aspects, in order to find situations or activities that may cause undesirable effects such as accidents, diseases or discomfort. The evaluation of adverse situations is also part of the assessment.

References and Further Reading 9

ABISOU G. De l'étable à l'étal. Les conditions de travail dans la filière viande. Montrouge: ANACT, 1993.

ALDRICH PT. Preventive services for health and safety at work in Denmark. Lyngby: Technical University of Denmark, 1991.

ALTMANN K. Betriebliches Modell zur Integration von Arbeitssicherheit, Gesundheits- und Umweltschutz, Qualitätsmanagement und Administratiever Organization. Bentheim: Deutsche Tiefborh-AG, 1993.

ANACT. Refaire "Le Monde". Montrouge: ANACT, La lettre d'information, mars, 1989.

ARENAZ ERBURU JC, GIL FISA A, HERNÁNDEZ CALLEJA A, LUNA MENDAZA P, NOGAREDA CUIXART S, PAREJA MALAGÓN F, PIQUÉ ARDANUY T, TURMO SIERRA E. Evaluación de las condiciones de trabajo en pequeñas y medianas empresas. Metodología práctica. Barcelona: Instituto Nacional de Condiciones de Trabajo, 1994.

BAILEY SR, JØRGENSEN K, KRÜGER W, LITSKE H. Economic incentives to improve the working environment. Dublin: European Foundation for the Improvement of Living and Working Conditions, 1994.

BERNDSEN M, VAAS, F Checklist for the evaluation of administrative work at computer screens. Dutch Trade Union Federation (FNV), TNO: Amsterdam, Leiden, 1991.

BG-BAU-Arbeitskreis "Gefährdungsanalyse". Anleitung zur Erstellung von Gefährdungskatalogen durch die Berufsgenossenschaften. Sankt Augustin: BG-BAU

BOSCH H. Risico-inventarisatie: fundament van preventief arbobeleid. Arbeidsomstandigheden, 1994. 70(1): 13-15.

DER BUNDESMINISTER FÜR ARBEIT UND SOZIALORDNUNG. Prävention im Betrieb. Arbeitsbedingungen gesundheitsgerecht gestalten. Bonn: Der Bundesminister für Arbeit und Sozialordnung, 1992.

BUNDESANTALT FÜR ARBEITSSCHUTZ. Specimen for a company health report. Official Journal of the Federal Institute for Occupational Safety and Health, July 1991. Dortmund: Bundesantalt für Arbeitsschutz, 1991.

BUNDESANTALT FÜR ARBEITSSCHUTZ. Grundmethodik Betriebliche Gefährdungsanalyse, December 1993. Dresden: Bundesantalt für Arbeitsschutz, 1993.

COX SJ, TAIT NRS. Reliability, Safety & Risk Management, An Integrated Approach. Oxford: Butterworths-Heinemann Ltd, 1991.

ELLENS E, BEUMER PFM. De Inspectiemethode Arbeidsomstandigheden. Zeist: Kerckebosch, 1992.

EUROPEAN FOUNDATION FOR THE IMPROVEMENT OF LIVING AND WORKING CONDITIONS. HASTE, European health and safety Database. Dublin: European Foundation for the Improvement of Living and Working Conditions, 1993.

FISCHER H, KIRCHBERG S. Erfordernisse und Methoden zum Erkennen und Beurteilen von Gefährdungen. Dortmund: Bundesanstallt für Arbeitsschutz, Tb60; 1993; 23-42.

GAMBA R, ABISOU G. La protection des travailleurs contre le bruit. Les points clés. Montrouge: Éditions de l'ANACT, 1992.

GUÉRIN F, ROUILLEAULT H. Du projet à l'evaluation de la conduite de projet. Communication au Symposium International sur la Productivité, Vancouver: ANACT, 1994.

HÄKKINEN KK. Improving industrial safety by self auditing. In Moncelon B (ed). Maitriser le risque au poste de travail. Actes du IVe colloque international du comité de recherches de l'Association de la Sécurité sociale. Nancy: Presses Universitaires de Nancy, 1993. 93-97.

HAUPTVERBAND DER GEWERBLICHEN BERUFSGENOSSENSCHAFTEN. GESTIS, Gefahrstoffinformationssystem der gewerblichen Berufsgenossenschaften, Aufgaben und Ziele. Sankt Augustin: Hauptverband der gewerblichen Berufsgenossenschaften e.V., 1990.

HEALTH & SAFETY EXECUTIVE. Successful Health and Safety Management. Sudbury: HSE Books, 1991.

HEALTH & SAFETY EXECUTIVE. Five Steps to Successful Health and Safety Management, special help for directors and managers. Sheffield: HSE Information Centre, 1992.

HEZIK HJPM van, ZWETSLOOT GIJM. Afstemming van eisen aan zorgsystemen voor kwaliteit, milieu en arbeidsomstandigheden. 's Hertogenbosch: NEHEM, 1994.

HOOR M, JANKE H, LOHRUM B, PUTZ H, SCHREURS B, STRASSER H. Gefährdungen und Schutzziele in Stahlwerken. Essen: Vulkan-Verlag, 1993

HUF C. Qualitätsanforderungen an den Arbeitsschutz- betriebliche Umsetzung. In:Qualitätsmanagement und Arbeitsschutz (Tb 64). Dortmund: Bundesanstalt für Arbeitsschutz, 1993.

INTERNATIONAL STANDARDS ORGANIZATION. Quality management and quality system elements - Guidelines, ISO 9004. ISO, 1987.

JENSEN PL. STRANDDORF, J MØLLER N. Developing safety management in practice. Lyngby: Institut for Arbejdsmiljø, 1993.

JOHANSSON B, JOHANSSON J. Working Environment and Internal Control in Small Companies. Luleå: Luleå University of Technology, 1993.

JÜRS H. Sichere Steurerung an halb/vollautomatische Anlagen mit modernen Rechnrsystemen. Moderne Unfallverhütung, Heft 38, 1994; 32-40.

KIRCHBERG S. FISCHER H. Erfordernisse und Methoden zum Erkennen und Beurteilen von Gefährdungen. In: Erkennen und Beurteilen von Gefährdungen. Dortmund: Bundesanstalt für Arbeitsschutz, Tb 60, 1993; 23-42.

KLOPPENBURG H. Guidance on Risk Assessment at work. EG/DG V/E/2 ref. HK/seh/7503 B. Luxembourg: DG-V, 1993.

KOMPIER MAJ, LEVI L. Stress at the workplace: causes, effects and prevention; guide for small and medium sized enterprises. Dublin: European Foundation for the Improvement of Living and Working Conditions, 1993.

KUHN K. POPPENDICK KE. Gefährdungsanalysen im Arbeitsschutz. In: Prävention im Betrieb, Arbeitsbedingungen gesundheitsgerecht gestalten. Bonn: Bundesminister für Arbeit und Socialordnung, 1992; 190-200.

KUHN K. Aufgaben der Betrieblichen ubd überbetrieblichen Arbeitsschutz - Organization bei der Vorbereitung und Durchführung von Gefährdungsanalysen. In: Erkennen und Beurteilen von Gefährdungen. Dortmund: Bundesanstalt für Arbeitsschutz, Tb 60, 1993; 220-234.

LANGE P. Methode zur Gefährdungsanalyse für elektrotechnische Anlagen und Betriebsmittel. Dresden: Bundesanstalt für Arbeitsschutz, 1994. S34

LIGTERINGEN B. Iedere werknemer is deskundig op zijn eigen werkplek. Arbomagazine, 1994, 10(2), 12-13.

LINDSAY FD. Succesful health and safety managemant. The contribution of management audit. Safety Science 1992;15:387-402.

LOHRUM B. Einbeziehung der Arbeitnehmer in die Gefährdungsanalysen. In: Erkennen und Beurteilen von Gefährdungen. Dortmund: Bundesanstalt für Arbeitsschutz, Tb 60, 1993; 235-258.

LORENT P. Europa voor veiligheid en gezondheid op het werk, veiligheid en gezondheid in de bouwnijverheid (Dutch version). Luxembourg: Bureau for official publications of the European Communities, 1993.

LOUGHBOROUGH UNIVERSITY OF TECHNOLOGY. Tackling Risk Assessment, a toolkit in support of the Management of Health and Safety at Work Regulations. Loughborough: Loughborough University of Technology, 1993.

MALINE J. Simuler le travail, une aide à la conduite de projet. Lyon: ANACT, 1994.

MARK HJ Van Der. Abstracte begrippen concreet maken, inventariseren van arbo-risico's. Gids voor Personeelsmanagement, 1993; 1, 36-39.

MEFFERT K. Abschätzung und Bewertung komplexer Risiken Klassification von Risiken und technischen Maßnahmen. In: Moncelon B (ed). Maitriser le risque au poste de travail. Actes du IVe colloque international du comité de recherches de l'Association de la Sécurité sociale. Nancy: Presses Universitaires de Nancy, 1993. 136-147.

MEULENBELD C, LINGEN P van. Instrumenten op het terrein van arbeid en gezondheid, een eerste inventarisatie. Amsterdam, Leiden: NIA, NIPG, 1993.

MEYER J, HARREN W. Die Management Strategie "Verbesserung des Sicherheitsbewußtseins als Unternehmungsziel". Heidelberg: Internationale Vereinigung für Soziale Sicherheit, 1993.

MINISTRY OF LOCAL GOVERNMENT. Internal Control, Regulations, with guidelines. Ministry of Local Government, Oslo, 1992.

MINISTRY OF SOCIAL AFFAIRS AND EMPLOYMENT. Arbo- en verzuimbeleid, P 190 (in Dutch). The Hague: Sdu, 1994.

MINISTÈRE DU TRAVAIL, DE L'EMPLOI ET DE LA FORMATION PROFESSIONELLE. Risques professionelles, guide d'evaluation, Paris: Ministère du Travail, de l'Emploi et de la Formation Professionelle, 1994.

MONCELON B (ed). Maitriser le risque au poste de travail. Actes du IVe colloque international du comité de recherches de l'Association de la Sécurité sociale. Nancy: Presses Universitaires de Nancy, 1993.

NOHL J, THIEMECKE H. Systematik zur durchführung van Gefährdungsanalysen, Teil I: Theoretische Grundlagen. Dortmund: Bundesanstalt für Arbeitsschutz, Fb 536; 1988.

NOHL J, THIEMECKE H. Systematik zur Durchführung von Gefährdungsanalysen, Teil II: Praxisbezogene Anwendung. Dortmund: Bundesanstalt für Arbeitsschutz, Fb 542; 1988.

NOHL J. Reduzierung von Unfällen durch vorausschauende Gefährdungsermittlung. In: Prävention im Betrieb, Arbeitsbedingungen gesundheitsgerecht gestalten. Bonn: Bundesminister für Arbeit und Socialordnung, 1992; 201-212.

OIRBONS JW. Tussen beheersing en verbetering. Op weg naar lerende, certificeerbare arbozorgsystemen. MSc Thesis. Tilburg: K.U. Brabant 1993.

OORTMANS GELLINGS P. Safety Audits, een verkenning. Arbeidsomstandigheden, 1990. 66(4):219-227.

PETERS AHMH. Quick-scan inventariseert risico 's bij DAF VGW-breed. Arbeidsomstandigheden, 1994. 70(3): 131-133.

POT FD, CHRISTIS JHP, FRUYTIER BGM, KOMMERS H, MIDDENDORP J PEETERS MHH, VAAS S. Outlines of the WEBA-instrument, a conditional approach for the assessment of the quality of work. Leiden: TNO, 1992.

PURNELL C. The impact of the COSHH Regulations on the working environment. Int. Journ. of Regulatory Law & Practice, 1992 (1): 205-214.

RASMUSSEN BH. Working environment management systems, trends, dilemmas, problems. In: Bradley GE, Hendrick HW (Eds) Human Factors in organizational design and management - IV. Amsterdam: Elsevier, 1994; 43-48.

RITTER A. Kleingruppenunterstützte Sicherheitsarbeit- Stand und Perspectiven. In: Erkennen und Beurteilen von Gefährdungen bei der Arbeit. Dortmund: Bundesanstalt für Arbeitsschutz, Tb 60; 1993. 261-275

ROUHIAINEN V. QUASA: A method for assessing the quality of safety analysis. Safety Science, 1992. 15, 155-172.

ROY O Du. The factory of the future. Dublin: European Foundation for the Improvement of Living and Working Conditions, 1990.

RUIGEWAARD PWG. Risico-inventarisatie- en evaluatie. Deventer: Kluwer Seminars, 1993.

RUPPERT F, HIRSCH C, WALDHERR B. Wahrnehmen und Erkennen von Gefahren am Arbeitsplatz. Dortmund: Bundesanstalt für Arbeitsschutz, Fb 426; 1985

SAARI J. Successful Implementation of Occupational Health and Safety Programs in Manufacturing for the 1990s. Int. Journ. of Human Factors in Manufacturing, 1992; 2(1): 55-66.

SAUNDERS R, WHEELER T. Handbook of Safety Management. London: Pitman, 1991.

SCHEIDERS H. Werkboek moet leiden tot arbojaarplan in midden- en kleinbedrijf metaal. Arbeidsomstandigheden, 1992; 68(5):291-295.

SCHÜTZ A, COENEN W. BIA-Handbuch Sicherheit und Gesundheitsschutz am Arbeitsplatz. Sankt Augustin: Berufsgenossenschaftliches Institut für Arbeitssicherheit, 1985.

SIMMONS S, ØLAND JS. Workplace Assessment. Dublin: European Foundation for the Improvement of Living and Working Conditions; 1992.

STEVENS TJ. Tien jaar risico-inventarisatie met het Inspectie Plus Pakket (IPP). Deventer, Kluwer Seminars, 1993.

SUTTER C, RACHEDI MF, MALINE J, GUERIN, F. Informer les salariés dans la conduite de projet. Montrouge: ANACT, La lettre d'information , janvier, 1992.

TCO. Resumé of TCO's Software Checker. Stockholm:TCO 1990.

TNO. SAFESPEC, Information and report system concerning working conditions. Apeldoorn: TNO, 1990.

THEFIOUX D. Le CHSCT: quel changements?. Montrouge: ANACT, 1993.

VOLKSWAGEN AG, GESUNDHEITSWESEN. Fragebogen zur Gesundheit und zum Wohlbefinden.

WORKERS' HEALTH INTERNATIONAL NEWSLETTER. When it comes to their health, workers always know best. Sheffield: Workers' Health International Newsletter, 1994, winter, 10-11.

WYNNE R, CLARKIN N. Under construction; building for health in the EC workplace. Dublin: European Foundation for the Improvement of Living and Working Conditions, 1992.

ZWAARD AW. De groenteboer hoeft geen foutenboom te maken Arbeidsomstandigheden, 1993; 69(1):7-10.

ZWAARD AW. Tien misvattingen over risico-inventarisatie. Arbeidsomstandigheden, 70(1): 17-20.

ZWETSLOOT GIJM. Joint management of working conditions, environment and quality. Amsterdam: NIA, 1994.

ØLAND JS Working Environment Management. Notes to the course "Consequences of Working Environment". Sorø: Industrial Health Service of Mid Zealand, 1993.

European Foundation for the Improvement of Living and Working Conditions

Assessing Working Conditions – **The European practice**

Luxembourg: Office for Official Publications of the European Communities

1996 – 126 pp. – 14.8 cm x 21 cm

ISBN 92-827-6098-7

Price (excluding VAT) in Luxembourg: ECU 11.50